Enclosed Waters

WAGENINGEN UNIVERSITY
WATER RESOURCES SERIES 9

This series published interdisciplinary analyses of water resources issues in Asia, particularly South Asia. The volumes are mostly Ph.D. theses defended at Wageningen University, the Netherlands, under the 'Matching Technology and Institutions' Ph.D. research programme that focusses on the interdisciplinary study of tank, well and canal irrigatin management in India and Nepal. The programme is implemented by the Irrigation and Water Engineering group and the Law and Governance group of Wageningen University, and is financially supported by the Ford Foundation.

Enclosed Waters

Property Rights, Technology and Ecology in the Management of Water Resources in Palakkad, Kerala

Jyothi Krishnan

Orient BlackSwan

ORIENT BLACKSWAN PRIVATE LIMITED

Registered Office
3-6-752 Himayatnagar, Hyderabad 500 029 (A.P.), India
Email: centraloffice@orientblackswan.com

Other Offices
Bangalore / Bhopal / Bhubaneshwar / Chennai
Ernakulam / Guwahati / Hyderabad / Jaipur
Kolkata / Lucknow / New Delhi / Patna

First Published by Orient Blackswan 2009

ISBN 978-81-250-3692-0

Printed at
Printline Printers
Hyderabad 500 029

Published by
Orient Blackswan Private Limited
3-6-752 Himayatnagar
Hyderabad 500 029
Email: hyderabad@orientblackswan.com

Contents

Tables

Figures

Maps

Boxes

Photos

All photos are by Abey George

Preface

The present academic endeavour provided me the opportunity to understand the many intricacies of land and water management. It has enhanced my understanding of how sustainability and equity are sidelined in day-to-day land and water management practices. A direct field understanding of the situation in a small part of the Palakkad district in Kerala, shows that change needs to come about at all levels, from the policy level to the individual level.

I am thankful to the Irrigation and Water Engineering, and the Law and Governance departments of the Wageningen University, as well as to the research programme, 'Matching Technology and Institutions in Land and Water Management' hosted by the above departments for providing me with an opportunity to pursue doctoral studies in this field. I thank Dr. Peter Mollinga, for his guidance throughout the study period. His prompt and sharp comments on each chapter draft have gone a long way in improving the quality of the final manuscript. I thank him for not giving up on me, despite a considerable delay in completion. I am grateful to Prof. Franz von Benda-Beckmann, for his continued support throughout the programme. Discussions with him have sharpened my understanding on the property rights dimension of natural resource management. His visit to the study area in 2002 helped me to focus my writing in a significant way. I also thank Franz for so warmly welcoming me to Halle, and in facilitating my search for literature while I was there. I thank Prof. Linden Vincent, for her critical and extremely helpful comments during the final stages of work, which helped in improving the quality of the final draft. I thank her for taking interest in my work and for helping me at a very critical stage. I take this opportunity to once again thank Franz, Linden and Peter for their sustained support and encouragement, especially at times when I thought I would not complete the research.

I take this opportunity to thank two other people, Dr. Sathis Chandran Nair and Dr. S. Santhi. Though they have not been directly involved in the present study, my interaction with them earlier on introduced me to the importance of studying land and water management, and to the importance of seeing and understanding the ecological dimension. I thank them for those first introductory sessions in the field during the few occasions when I was fortunate enough to do field work with them.

I thank the farmers from Kollengode and Elavenchery, whom I met with during the period of fieldwork. They welcomed us to their homes and explained to me the intricacies of paddy cultivation and pond management in the area. I have repeatedly visited many of them, in order to clarify many doubts and confusions. I thank them for their patience and warm hospitality.

I am thankful to libraries in both India and Wageningen for facilitating the search for relevant material. I thank the library staff at the Leeuwenborch library at Wageningen who did all that was possible to get me all the books I needed during my short one-month stay there in 2006. The online access to journals provided by the Wageningen University, accessible even outside the Netherlands, was of big help. In India, I am thankful to the libraries at Centre for Development Studies, Trivandrum; the Kerala Forest Research Institute (KFRI), Peechi, and the Institute of Socio-Economic Change, Bangalore. At ISEC, Esha Shah and Ganesha were very helpful. I also thank Vellingiri who helped me with library work at CDS.

I thank Mr R Jayarajan and Ms. Kadeeja, for working so hard on the maps in the book. I also thank Ms. Bineetha for critical help in the final stages of formatting the text. I am also thankful to Veenu Luthria from Orient BlackSwan for her kind help in coordinating and helping out with the final nitty-gritty's of editing and formatting. I thank Ms. Rukmini for the discussions I have had with her on the issue of water use in agriculture. I thank Dr J B Rajan for reading through the final draft and suggesting improvisations. I also thank Ms. Selvi and Ms. Ajitha for assisting with field data collection and data entry.

At Wageningen I am thankful to interactions with researchers and scholars in both the IWE and LAG departments. Special thanks to Rutgerd Boelens, Dik Roth and Margreet Zwarteveen in this regard. I thank Dik Roth for his continued encouragement and support, and for making my second stay at Wageningen as fruitful as possible. In particular, I thank him for so very kindly translating the thesis summary into Dutch, despite many other engagements. I thank Maria Pierce, Gerda de Fauw and Trudy Freriks from the Irrigation and Water Engineering department, and Ellen Wegkamp and Maarit Junnikkala from the Law and Governance department, for helping out with many administrative matters. I especially thank Maria Pierce for being so very helpful during my (and my family's) short stay at Wageningen in 2006, and Maarit Junnikkala for her kind and thoughtful ways during the period of thesis submission and defence. I thank fellow travellers in the MTI group, R Manimohan, Diana Suhardiman, Bala Raju Nikku, Esha Shah, Vishal Narain, Suman Gautam, Anjal Prakash, Preeta Lall and Pushpa Raj Khanal. The MTI workshops I attended were very helpful in reviewing and improving upon my work.

Finally I thank my family and friends, for their support and encouragement in numerous ways. Being a mother of two very small children, this work could not have been completed without their support. First of all, I thank my parents, Mr R. Unnikrishnan and Ms. Jaya Unnikrishnan, for supporting all my academic endeavours. I thank them for letting me study the subjects of my choice and for supporting it in every possible way. I thank my father for inculcating in me the habit of disciplined note making at a very young age, which has come of immense help, while reading and writing. I also thank him for meticulously reading through the final draft and making corrections. During the last two years of thesis writing, my parents' help was extremely critical, as they took care of our little children while I worked. My mother has put aside all her schedules in order to be with the children. I thank my brother Pradeep and sister-in-law Reshmi too for their support and encouragement. I also take this opportunity to thank my uncle Mr V A Menon for introducing me to the importance of doing socially relevant research. His commitment to such research, despite all odds, is a big source of inspiration.

I feel fortunate to have parents in laws who have been an immense source of support and strength throughout the study period. I thank Mr K M George and Ms Thressiamma George and my sister-in-law Bindu for their support and prayers. I am also grateful to them for so graciously accommodating our long absence from home during this period.

At Wageningen, I thank my dear friends, Margreet Bossen and Geertje Rezelman, who made Wageningen home for me. I treasure the time I spent with them, at their home and working in their vegetable garden. Through them I have been able to learn a lot more about Dutch society as well. I also thank Hugo and Angela, for their warmth and friendship. And of course for assisting with bicycle repairs too, which turned out to be as serious as doing a PhD! I also thank the Gleichman family at Belmontelaan for their kind hospitality during our one-month stay at Wageningen in 2006.

While doing fieldwork at Kollengode, I am indebted to the Prabhakaran family at Palakode for taking care of our little son with so much of love and care. Fieldwork would have been very difficult without their support, in particular the support of Prema edathi and Nitya. I also thank Suneeta, Sumati and Rosy chedathi for lovingly taking care of our children at different stages of work. I also take this opportunity to thank friends for their re-assurance in many ways. To Kochu (Anitha) for not letting me lose touch with larger quests and dreams, Kali, Sival, Sohini, Bindu and Vini. During the last year of intensive writing, I was fortunate to receive the help of two families. I thank Mr Y. Sreekumar and Ms Radha Sreekumar for so kindly providing me with a room in their house, where I could read and write peacefully. I also thank Rajan and Bindu for letting me invade their house with my books and files, and for providing a second home for our children.

Finally, I thank my friend and companion Abey George, for his unfaltering support from beginning to end. His understanding of issues and experience in this field has contributed in a big way towards this research. Right from the stage of proposal writing, to doing fieldwork and discussing chapter drafts, his support has been crucial, helping me to take the work ahead, keeping me away from giving it up at difficult moments. His spirited approach and commitment to doing socially relevant and meaningful research continues to be my biggest source of strength and inspiration.

And I thank our children, Adityan and Arunima, for bearing with it all and for brightening our days. I am relieved to have finally answered my son's persistent question, 'Amma, when will you be over with your PhD?'.

Glossary of Local Terms

Adharam	Title deed
Adi adharam	Parent title deed for a piece of land
Aela(s)	Meandering low valleys
Ahars	Traditional water harvesting structures found in the state of Bihar
Asari	Carpenters
Ayacut	Command Area
Block Panchayat	The second tier in the three-tier local governance system in the country
Brahmins	A Hindu higher caste group
Chaaku	Sack
Chaals	Drainage channels
Chama	Little millet
Chera	Paddy fields cultivated with paddy only during the first crop season. During the second crop, they are stocked with water.
Cheru	Silt that accumulates on the pond bed
Cherumakkal	A Scheduled Caste community who have been traditionally engaged as agricultural labourers
Chettiar	Tamil speaking caste group in the region
Chetu veda	Broadcasting of paddy seeds in slush
Cholam	Corn
Chungam	Toll
Desam	A territorial division of the past
Eri	Large pond
Ezhavas	A Hindu caste group in Kerala

Gramams	Housing cluster inhabited by the Tamil Brahmins
Jalanidhi	A government programme to ensure rural water supply and sanitation
Janmi	Landlord
Kalam/Kalapera	Farm house
Kalayi	Low lying, double cropped paddy fields
Kali valam	Fertiliser comprised primarily of cow dung and other bio waste
Kanakar	Principal tenants
Kanam	Tenancy rights of the principal tenants.
Kanni	A month in the Malayalam calendar, which coincides with the months of September-October in the English calendar. It is also the month when the first crop is harvested, and hence the first crop is often referred to as the Kanni crop.
Karyasthan	Supervisor appointed by the landlord
Kavu	Place of worship
Kazhayi	Open mud valve located on the bund of paddy fields to regulate the inflow and outflow of water
Khadins	Traditional water harvesting structures found in the state of Rajasthan
Kollan	Blacksmiths
Kollinyal	A leguminous variety used as mulch
Koravar	Bamboo artisans
Krishi Bhavan	The Agricultural Office of each panchayat
Kudiyan	Generic term referring to tenants in general
Kudiyirippu	Housing cluster inhabited by agricultural labourers
Kulam	Pond
Kumbaran	Potter community
Kuzhis	Shallow pits in paddy fields
Maestry	Supervisor of water distribution in the canal network

Mele kandam/ Metu kandam	Paddy fields located on elevated land
Mele ovu	Higher sluice
Michabhoomi	Land in excess of the stipulated ceiling that had to be surrendered by the concerned land owners after the implementation of land reforms in the state
Naatu Potta	Fields on which saplings for the second crop of paddy are raised.
Naatukootam	Gathering of local people
Nairs	A Hindu caste group in Kerala
Namboodiri	A Hindu higher caste group in Kerala
Nanavu	Wetness
Nelkrishi vikasana agency	Paddy Development Board
Nilom	Levelled paddy fields
Onam	A state festival celebrated after the harvest of the first paddy crop, in the months of August-September.
Ootu	Sub-surface drainage of water to the low lying paddy fields
Ovu	Pond Sluice
Paambum kavu	Place of worship where the snake-god is the deity
Pachila valam	Fertiliser made out of leaf litter
Padashekhara samity	Registered body of paddy farmers
Pana	Palmyra
Panayude pati	Pipe like structure obtained by scooping out the insides of the bark of the Palymra tree
Panchayat (Grama Panchayat)	The lowest tier in the three-tier local governance system in the country
Pandaranmaar	A community which migrated into the region from Andhra Pradesh
Para	A local measure of grain

Parambu	Land located on the higher slopes, partly occupied with houses and a wide mix of trees.
Parayar	A community who were traditionally engaged in desilting of ponds
Patam	Rent paid by the tenant to the landlord in the pre land reform era.
Pattar	Tamil Brahmins
Podi veda	Broadcasting of paddy seeds in dry soil
Potta	Single cropped paddy land
Pudu vellam	The first showers of rain
Puli avara	A variety of beans
Puzha	River
Ragi	Finger millet
Sabha	Gathering
Sarkar vellam	Government's water, referring to water supplied through the government owned canal network
Sthira panikkaar	Permanent agricultural workers
Swarupam	Small kingdom
Talakulam	Pond at the head of the slope
Taluk(s)	An administrative division
Tanapu	Cool
Tara	Housing cluster
Tazhe ovu	Lower sluice
Valartukadu	Pocket of dense vegetation
Varagu	Proso millet
Varambu	Pond bund
Verumpattakaran	Sub tenant
Verumpattam	Rent paid by the sub tenant
Vetti	Basket
Vikasana rekha	Development Plans prepared by each Panchayat in the state
Vishu	Vishu is the astronomical New Year day festival celebrated on April 14th in the state of Kerala, coinciding with the first day of the Malayalam month of Medam.

Vishu kaineetam	A token sum of money gifted by the elders to the younger members of a family, and by employers to employees on the occasion of Vishu
Vrichikam	A month in the Malayalam calendar, which coincides with the months of November and December in the English calendar.
Vrichikapandi	Name of a paddy seed, which is harvested in the Malayalam month of Vrichikam.
Zilla Panchayat	The District Panchayat, which represents the third tier in the three-tier local governance system in the country.

Abbreviations

BFA	Beneficiary Farmer Association
CADA	Command Area Development Authority
cu.m	Cubic metres
CWRDM	Centre for Water Resources Development and Management
DPC	District Planning Committee
GALASA	Group Approach for Locally Adapted and Sustainable Agriculture Scheme
GOI	Government of India
GOK	Government of Kerala
GWP	Global Water Partnership
Ha	Hectare
HP	Horse Power
HYV	High-Yielding Variety
IADP	Integrated Area Development Programme
IPD	Integrated Paddy Development
IWRM	Integrated Water Resource Management
JCB	Joseph Cyril Bamford (name of the person who started the company JCB that manufactures heavy agricultural equipment).
KAU	Kerala Agricultural University
Kg	Kilogram
Km	Kilometres
KSEB	Kerala State Electricity Board
KSI	Kerala Statistical Institute
LBC	Left Bank Canal
mcft	Million cubic feet
MLA	Member of Legislative Assembly

mm	Millimetres
MP	Member of Parliament
NABARD	National Bank for Agriculture and Rural Development
O & M	Operation and Management
OBC	Other Backward Caste
PAC	Project Advisory Committee
PAP	Parambikulam-Aliyar Project
PIM	Participatory Irrigation Management
SC	Scheduled Caste
SPB	State Planning Board
sq.km	Square kilometre
tmc ft	Thousand million cubic feet
UN	United Nations
UNDP	United Nations Development Programme
WCD	World Commission on Dams

1

Introduction

This study commenced as an enquiry into the phenomenon of water scarcity in the southeastern parts of Palakkad district in the state of Kerala, in south India. The district of Palakkad, most parts of which drain into the Bharatha*puzha*[1] River, is considered to be one of the rice bowls of the state of Kerala.[2] Much prior to the introduction of modern irrigation, the Palakkad plains were distinguished as a paddy-producing region (Innes 1908). Widespread deforestation, changing land use patterns, transformations in the management and use of water resources and the process of agricultural intensification have been some of the hallmarks of the changes that have afflicted the region over the past four to five decades. Palakkad today is known more for the acute water scarcity and accelerating desertification it experiences (S. C. Nair 2004).

While the district continues to top the state in terms of acreage under paddy cultivation and total production of paddy,[3] it is has registered a significant decline on both these fronts (see Chapter 2).

The existing irrigation infrastructure is also in a state of disrepair. Newspaper reports of water scarcity[4] during every winter crop[5] of paddy narrating the woes of farmers, depicting the sad sight of parched paddy fields, empty reservoirs and neglected ponds (*kulams/eris*) have become a routine affair.

Water scarcity is manifest in the inadequate supply of water for drinking and irrigation purposes in many parts of the district. The government has had to provide compensation to paddy farmers who have faced crop losses owing to inadequate water supplies, during the winter crop season on a number of occasions in the recent past. This has been particularly so in the case of the specific study area (the *panchayats*[6] of Kollengode and Elavenchery)[7] in the southeastern *taluk*[8] of Chittur in the district. The unique agro climatic features of the southern and eastern parts of Palakkad district (characterized by relatively lower amounts of rainfall and higher temperatures when compared with the state average) has however contributed to the popular perception that water scarcity is 'but natural' in such a context. Preliminary rounds of fieldwork in the area indicated that water scarcity was not as natural as it was made out to be. The issue of water scarcity raised important questions regarding the management and use of water and other natural resources in the area.

The increasing severity of the problem of water scarcity across the world has prompted much research on this topic. Initial research was focused on illustrating the magnitude of the problem in quantitative terms, arguing that there was not enough water for an increasing population (Gleick 1993, Falkenmark and Widstrand 1992). It has been subsequently pointed out that while estimates of the population-water equation (UNDP 2006), is an important indicator of the impending crisis, it provides only a part of the picture (Falkenmark and Lundqvist 1998). Increasing attention is now being paid to the management and distributional aspects of water resources management and its link with the water crisis faced in many parts of the world today. The Human Development Report of 2006 for instance points out that 'much of what passes for scarcity is a policy-induced consequence of mismanaging water resources' (UNDP 2006). This has been attributed to the prevailing view that water is there to be exploited without reference to environmental sustainability (ibid.). Similarly on the distribution front, the role of socio-political and institutional processes have been highlighted in order to counter the tendency to 'naturalise' the phenomenon of water scarcity (Mehta 2007). Mehta argues that the story of less rainfall in many a water scarce region obscures the intricate web of power and social relations that govern access to and control over water (ibid.). Over-allocation and use of rivers and aquifers coupled with unequal access have therefore been pointed out as important parameters in the study of water scarcity (Vincent 2004). Equally important is the decay of irrigation infrastructure often due to poor management practices and the failure to create supportive institutional arrangements to govern water supplies (Mehta 2003), a condition that is referred to as managerial and institutional scarcity (Molle and Mollinga 2003).

What began as an enquiry into the problem of water scarcity grew into a study of the socio-political, institutional and hydrological factors that determine the availability and distribution of water. This study has illustrated the significant lack of concern for the goals of sustainability and equity in the institutional mechanisms that guide the management and distribution of the water resources of the study area. It shows how the existing state level irrigation and agricultural policies, by patronizing a single crop (paddy) and a single mode of irrigation have paid little attention to local topographical and agro climatic features that influence the availability and use of water in each region. In the study area, this led to the neglect of the pond based agricultural and irrigation practices. It also shows how the existing water resources 'development' approach, by focusing on infrastructure oriented solutions neglects the long-term sustainability of locally available water resources. With regard to the distribution of existing water supplies, this book finds the predominantly private property mode of resource appropriation unsuitable to an equitable distribution of water. The existing formulation of property rights over land and water has also been critiqued for its disregard for the ecological characteristics of the resource. It sanctions a pattern of resource use that is unsuited to the demands of long-term ecological sustainability.

This chapter is divided into four sections. The first section outlines the conceptual contours that have guided this enquiry, followed by the central research question. The second section introduces the changing ecological and social landscape in which the study has been conducted. The third section discusses the methodological design adopted and the fourth section provides an outline of the structure of the book.

Central Research Question

The central research question is framed as follows:

How does the interaction between the ecological, institutional and technological dimensions of the prevailing mode of water resources management and distribution explain the manifestations of unsustainability and inequity? How does this interaction shape the problem of water scarcity in the study area?

The primary research concern of the present study is therefore with the relationship between the ecological, technological and institutional dimensions of water resource use and management, and its implications on sustainability and equity in the management and use of water. While doing so, the primary focus has been on the institutional aspects (in particular the existing property rights regime, the governance mechanisms, and the overall policy framework), and the extent to which they address the concerns of sustainability and equity.

The book focuses on two main issues, the management and the distribution of water resources. While discussing the underlying approach towards the management of water resources, it refers to the existing critiques of the conventional approach to water resources management that point to the neglect of the ecological dimension. It also shows how the prevailing supply oriented approach towards water resources management frames the problem of scarcity in supply terms, which neglects vital causative factors in the precipitation of scarcity.

The issue of equitable distribution of water resources is located within the property rights framework. This book illustrates how the creation of public and private rights over a fluid and common pool resource such as water results in skewed access to the resource. The situation is further aggravated by the operation of unequal power relations and the differential ability to invest in modern water extraction devices.

Conceptual Contours

The present section discusses concepts that help in undertaking an inter-disciplinary study of the ecological, institutional and technological dimensions of water resources management in order to provide an explanation for the problematic water scenario of the study area. The focus of the study is on analysing the factors that give rise to the present style of water resources management and distribution, and its implications on the phenomenon of water scarcity. It focuses in particular on the extent to which the institutional mechanisms that govern the management and use of water take cognisance of the specific ecological properties of water and how they shape the sustainability and equity (or the lack of it) in the management and distribution of water resources. This interaction between the institutional and ecological dimensions is mediated by the diverse technologies that make available water for irrigation, viz. the modern reservoir based canal technology, the traditional pond technology, the open wells and energised wells.

While referring to the specific ecological properties of water, the study refers to the fact that water, as a resource exists in relation to other natural resources like land and forests in its movement through the hydrological cycle. Water therefore is a resource that is not only in constant movement, but also in constant interaction with other resources. There also exists a constant interaction between different components of the hydrological cycle, such as surface and groundwater. These comprise the basic ecological properties of water as a resource that are often ignored while designing institutional mechanisms that govern the management and use of water on a daily basis.

While I discuss the relevance of the ecological dimensions of water management, I must state that I am not a trained ecologist. This research therefore does not deal with the complex ecological dynamics between land, water and forests or between surface water and groundwater. My attempt through this study has been to build a bridge between the social and ecological issues that pertain to water resources management. Ecological scientists may be disappointed by the inability of the study to explore in depth many of the ecological parameters (dealt with in Chapter 3) like the impact of deforestation on surface flows, or the extent of groundwater decline, or the hydrological impact of excessive extraction of surface and groundwater. While the study addresses ecological phenomena such as deforestation, changing land use patterns, dwindling surface flows and increasing groundwater extraction, it focuses on the social and institutional factors that have given rise to them.

With regard to the institutional mechanisms that govern the management and use of water, this study refers primarily to the institution of property rights over land and water, as well as the existing systems of governance, and their approach to the management and distribution of water resources in the study area. While discussing the former aspect, the existing rights regime and its impact on the distribution of water are taken up. Also discussed are the ambiguities that arise while categorising certain waters as public and certain others as private, which come in the way of an equitable distribution of existing water supplies in the study area. The book argues that the existing formulation of property rights over land and water needs to be cognisant of the ecological properties of a fluid resource such as water. The impact of the land reforms on sustainable and equitable land and water management is taken as a case in point. The book also analyses the extent to which different components of the governance set up (laws, policies, government departments, institutions of local self government, local level organisations and associations) address the issue of sustainability and equity. The recently implemented decentralisation reforms are evaluated from this angle.

Since the 1990s, a large number of inter-disciplinary studies in irrigation have been conducted (Bottrall 1992, Vincent 1995, Jurriens and Mollinga 1996, Mollinga 1998, Bolding 2004), wherein irrigation was viewed as both a social and technical phenomenon. The Matching Technology and Institutions Programme of the Irrigation and Water Engineering and the Law and Governance Departments at the Wageningen University was constituted to encourage research along these lines. Research conducted as a part of this programme illustrated the importance of the social dimensions in a field that was largely considered to be a technical one. The present study extends this inter disciplinary approach to include the ecological dimensions as well. As a result, it is not a particular irrigation system/s that has been focused upon, but the functioning of diverse irrigation systems located within the natural boundaries of a watershed.

Irrigation and the Sustainability of Water Resources

Irrigation since the second half of the past century has been focused on making available water for agriculture, through the use of capital-intensive technology. This supply-oriented approach to irrigation was the product of an era in which concerns about the finiteness of water resources were yet to be recognised. Referring to existing critiques in this regard, I point out some of the ecological concerns that have been by-passed in the hurry to enhance the supply of water for human consumption. I also argue that the existing fragmented approach towards the management and use of natural resources needs to be replaced by a more integrated one.

The existing water development culture across the world has been based on the mobilization and appropriation of water with an inherent resource capture dimension to it (Turton et al. 2001). Such an approach has been attributed to the reductionist perspective of traditional engineering, wherein water has been viewed as a stock of resource, to be withdrawn and utilised as desired (Bandyopadhyay 2006). The prevailing view of water resources is one in which water is considered to be infinitely available, to be diverted, drained and polluted (UNDP 2006). The emphasis has been on making available larger amounts of water for human consumption with the aid of capital and energy intensive technology, which is mostly referred to as 'development' of water resources. The predominance of such an engineering paradigm in modern irrigation development has led to an almost exclusive focus on infrastructure development. Molle describes this as a technocratic vision of water resource development (Molle 2005). While such a strategy has been justified by the increase in the number of hectares of land irrigated and infrastructure created, off late it has been critiqued for its neglect of the ecological transformations precipitated in the process (ibid.).

Such an approach to water resource management is an outcome of the industrial mode of resource use that has come to prevail largely since the Industrial Revolution. This mode of resource use endorses the view that nature and natural resources are mere commodities, at the disposal of human beings. Water was therefore viewed as a commodity to be made available through pumps, pipes and canals, disconnected from resources such as land and forests. Such a view of water obscured the ecological context of water from the mind of the user as well.

Water resource planning in most countries has been primarily equated with irrigation development, which has been largely centred on the construction of large dam and reservoir projects (Iyer 1998: 3198). In the case of groundwater, the development of mechanized drilling extended the reach of technology into the hitherto unreachable underground aquifers leading to a sudden spurt in groundwater irrigation (Shah and Roy 2002). All through, issues related to the sustainability of the water resource (be it a river, a stream, a pond or an underground aquifer) that feeds these irrigation systems have been sidelined.

As a result the high environmental costs of such a development trajectory manifest in aquatic ecosystem degradation, fragmentation and desiccation of rivers, drying up of wetlands (Molden and Fraiture 2004: 4) and the depletion of groundwater have been recognised only of late (Burke et al. 1999). The ecological impacts of dams on downstream river flows, on river morphology and the quality of river water has been a subject of detailed investigation in the recent past (Postel 1999:119, Iyer 1998, WCD 2000)[9]. Likewise, the ecological and economic consequences of intensive groundwater extraction manifest in aquifer mining, declining water tables, declining water quality, land subsidence and rising pumping costs have become too serious and widespread to be ignored (Molle et al. 2003: 14).

Recognition of these ecological implications has strengthened the argument in favour of a 'resource management' rather than a 'resource development' approach to water (Davidson and Stratford 2006, Roy and Shah 2002), indicating a recognition of the limits to extraction and appropriation of water for human consumption, and the need to manage the resource in a manner that ensures its long term sustainability. The formulation of the Dublin Principles in 1992 and the concept of 'environmental flows' follow this line of thought, emphasising that there are limits to the human appropriation of the hydrological cycle (GWP 2000). Since the formulation of the Dublin Principles there has been an explicit focus on ensuring sustainability and equity in water resources management. The UN 2000 Millennium Assembly emphasised conservation and stewardship in protecting our common environment, stressing the need to stop the unsustainable exploitation of water resources, by developing water management strategies which promote both equitable access and adequate supplies at the regional, national and local levels (Rogers and Hall 2003).

Thus we see a gradual shift in the approach towards the management of water resources. From an exclusive focus on technology, there is now greater emphasis on the resource per se. There is also a growing appreciation for an integrated view of resources and the need to view water in relation to other resources such as land and forests. This is manifest in the widespread support within academic and policy circles for the concept of integrated water resource management (GWP 2000, Mitchell 2005), which asks for a recognition of the biophysical linkages between different components of the hydrological cycle (Hufschmidt and Tejwani 1993). Recognizing that land use affects the quantity and quality of renewable freshwater resources, this approach argues that land and water should be managed in a coordinated and integrated manner (GWP 2000).

Such changes, however, have been largely confined to the policy level, with very little of these ideas being put into practice. In India for instance, the concept of integrated water resources management has not been implemented in any significant manner (Mollinga 2006).

Irrigation policy in Kerala

The state level irrigation policy reflects the infrastructure bias discussed above. Kerala's irrigation strategy has been centred on developing surface water resources with emphasis on the development of major and medium irrigation projects (GOK 1999). Out of a cumulative investment of 38810.6 million rupees in the irrigation sector until March 2006, 70.60% (Rupees 27401.3 million) was spent on major and medium irrigation projects (GOK 2007b). Investment of such a high order however could not generate a commensurate increase in area irrigated, with most projects remaining incomplete, resulting in cost escalations and time over runs (GOK 2003). Since the Seventh Plan period (1987–1992), the state has been giving increased emphasis on minor irrigation development[10], defined as irrigation works having a cultivable command area less than 2000 ha (GOK 2006d). Major and minor irrigation schemes have also been defined in terms of capital investment, with the former comprising of schemes costing over Rupees 1.5 million and the latter costing below the same amount (Kuruvilla 1967). Minor irrigation focuses on the construction of check dams, vented crossbars, weirs, tanks, lift irrigation schemes and on groundwater irrigation. The increasing emphasis on minor irrigation has not however been in response to the negative ecological effects of large-scale irrigation. To the contrary, the focus on minor irrigation has been the result of the constraints faced in the further development of major irrigation potential. For one, the financial constraints faced by governments at the Centre and the State have resulted in a reduction in public investment in major irrigation (GOK 2006d)[11]. In addition, possibilities for the creation of irrigation potential have been more or less exhausted, with the remaining potential sites being located in environmentally sensitive regions (GOK 2003). The more important reason behind the enhanced emphasis on minor irrigation has been the changing cropping patterns in the state, reflected in the reduction in the area under paddy and the increasing area under perennial crops such as coconut, areca nut, rubber, pepper and banana. It was therefore felt that large-scale irrigation projects that were designed to meet the seasonal water requirements of paddy cultivation were unsuitable to the year round water requirements of perennial crops. Hence, tapping the wide network of streams and rivers through small structures such as check dams and lift irrigation schemes was considered a more suitable strategy (GOK 2002a).

This shift in irrigation strategy has not changed the essentially extraction oriented approach towards water resources management. While rainwater harvesting and enhancing groundwater recharge have received some amount of emphasis during the Ninth and Tenth Five year plans (i.e. since 1997), they have not redefined the basic approach towards irrigation in order to incorporate a greater concern for the sustainability dimension.

Governance Reforms in Water Resources Management

An important rationale for a reconsideration of the predominant supply augmentation strategy in water resources planning is the fact that the period of rapid expansion of irrigated area through construction of surface irrigation systems or exploitation of groundwater is slowly nearing an end in most parts of the world (Barker and Molle 2004). At the policy level therefore, during the past decade and a half, efficient use of the existing water systems has been considered as an important strategy to ease the pressure on diminishing water supplies (Barker and Molle 2004, Cotula, 2006). This has led to a focus on 'non structural solutions' (WCD 2000), with greater attention being paid to the issue of governance and its role in ensuring efficient management of water supplies. Such a shift in thinking has also been strengthened by the growing emphasis on integrated water resources management (IWRM) as a means to ensure equitable, economically sound and environmentally sustainable management of water resources (Rogers and Hall 2003). By emphasizing the 'coordinated development and management of water, land and related resources' in such a way that neither equity nor sustainability of the ecosystems is compromised (GWP 2000), the IWRM philosophy puts pressure on existing governance systems to change their approach towards the management of water. The role of local governance mechanisms has received particular emphasis, with local forms of management considered to ensure sustainable and equitable water resources management (Benda-Beckmann and Benda-Beckmann 2003).

In parallel, the recognition of the ecological dimension in water resources management has prompted some to look at governance with a focus on the match and mismatch between the politico-administrative system and the ecological system (Rogers and Hall 2003). Gatzweiler and Hagedorn for instance argue that in order to create institutions of sustainability, the requirements of ecosystems function management needs to be matched with adequate property rights regimes and governance structures and arrangements (Gatzweiler and Hagedorn 2002). This match has been sought by adopting hydrological basins or catchment units, rather than administrative units as the basis for planning and implementation (Orindi and Huggins 2005). This has resulted in a growing trend in favour of forming governance institutions at the level of hydrological units, viz. around an irrigation system, a catchment area or a river basin (Donahue and Johnston 1998 in Benda-Beckmann and Benda-Beckmann 2003). Ostrom argues that such measures ensure a better match between institutions and the physical, biological and cultural environments in which they function (Ostrom 1990). This marks an important shift as far as conventional governance is concerned, as it demands an enhanced appreciation of the ecological aspects.

Against this background, this study analyses the extent to which local governance over water management in the study area has brought about changes in the prevailing infrastructure and supply oriented mode of water resources management. Through the ongoing decentralisation reforms in the state of Kerala, institutions of local self-government (the panchayats) have been vested with enhanced powers over natural resource management. In the case of water specifically, panchayats have been vested with powers to formulate local water resources management plans with regard to the water resources located within their boundaries. The experience so far has not been very encouraging. The focus continues to be on measures that enhance the extraction of water from existing sources, primarily through check dam cum lift irrigation schemes. Devolution of powers has not led to a change in the underlying approach towards water resources management, which continues to be viewed as an activity that enhances the supply of water for human consumption.

It is yet to be conceptualised as an activity that involves the integrated management of land and water resources.

The Scarcity Debate

Scarcity of water, as with the case of other natural resources, has been mostly viewed as an outcome of the increasing demands of a growing human population. Water scarcity has therefore been mostly captured in physical or volumetric terms with water stress indices being developed to indicate the level of scarcity (Shiklomanov 1997, Falkenmark and Widstrand 1992). More recent research on this topic has however made attempts to sieve out other causative factors, in particular the social and ecological dimensions of the phenomenon of scarcity.

It has for instance been argued that while certain regions are more vulnerable due to a shortage of naturally occurring water endowments, scarcity is more often the result of inadequate management of water resources (UNDP 2006). It has therefore been argued that water scarcity is not natural, but largely due to anthropogenic interventions in the realm of land and water management and use (Mehta 2003). Scarcity has also been viewed as an outcome of the underlying instrumentalist approach towards natural resources on the whole. The Human Development Report of 2006 for instance comments upon policy induced mismanagement of water resources, noting that public policies that sacrifice entire ecosystems for the cultivation of water hungry crops and those which sanction levels of water consumption that exceed the availability of renewable water resources are much to blame (UNDP 2006). Over-allocation and improperly controlled extraction of water (Vincent 2004) are therefore considered to play an important role in the exacerbation of physical scarcity.

Another important strand in the contemporary debate on water scarcity deals with the relational aspects of scarcity, i.e. primarily with the distribution of existing water supplies. Mehta's work illustrates how the powerful discourses of scarcity attempt to 'naturalise' it, obscuring the fact that access to and control over land and water resources is highly unequal (Mehta 2007). This is also referred to as 'resource capture', a process by which the powerful social groups shift resource distribution in their favour (Homer-Dixon and Percival 1996 in Turton and Ohlsson 1999). Such processes of resource capture have been argued to institutionalise privilege, becoming a source of structural scarcity (Turton and Ohlsson 1996). Emerging research has also conceived of scarcity in orders, delineating the physical from the institutional dimensions. Turton and Ohlsson have conceived of different orders of scarcity, with a first order emphasizing the physical dimension of scarcity (i.e. scarcity of the resource per se) and a second order emphasizing a scarcity of social resources (i.e. the ability and skills) to deal with physical scarcity (Turton and Ohlsson 1999, Vincent 2004). The latter essentially refers to the technological or institutional measures that make management of a natural resource more efficient (Wolfe and Brooks 2003). A third order has also been identified that emphasises the changes that a society must undertake to deal with scarcity, through education, cultural changes, re-evaluation of lifestyles and adopting less water intensive consumption patterns (ibid.). This book does not analyse these emerging frameworks in detail. It illustrates, however, that the present framing of the scarcity problem in the area ignores both ecological and socio-political realities. Three issues that have been consistently sidelined are the physical state of art of the resource, increasing demand, and unequal access to and control over the resource.

The book shows how the framing of scarcity in supply terms alone sidelines the degradation of the local water based ecosystems, viz. the streams and ponds. Factors that contribute to declining flow in streams and rivers, as well as declining storages in ponds and reservoirs have been sidelined while devising ameliorative measures. It is being acknowledged that in regions that depend on short rainy seasons, real availability of water is dependent not only on rainfall, but also on the capacity for storage and the degree to which river flows and groundwaters are replenished (UNDP 2006). Soil and water conservation, restoration of vegetative cover, wasteland regeneration and replenishment of groundwater reserves are some of the measures that can enhance availability of water (Torori et al. 1996, Mehta 2007). These measures however compare poorly with the quantum of water made available through inter basin transfers, with the balance tilting in favour of the latter. Scarcity redressal in the study area is focused on further supply augmentation through water diversion projects such as the Palakapandi and the Kuriarkutty-Karappara. Since the 1990s when the supply of canal water has become increasingly unreliable, the proposed Palakapandi project (now under implementation) in particular has been portrayed as the only lasting solution, by bureaucrats and politicians alike. Such an approach is in line with the dominant paradigm of developing infrastructure to supply more water to abate conditions of scarcity in the country (Vincent 2003: 2, Mehta 2003, Mahajan and Bhardwaj 2002). This book also shows that the reliance on engineering solutions distances the water user from the risk of over exploiting and degrading locally available surface and groundwater resources (Burke 2000). This feeds into a vicious cycle whereby the degradation of local water resources is not considered important in the manifestation of scarcity, as a result of which ameliorative measures continue to be focused on further supply augmentation.

The study also illustrates how an emphasis on structural measures to abate water scarcity detracts attention from unequal access to the existing water supplies, as well as from the issue of increasing consumption of water. Water use in the area is dictated by an increasingly private property regime over land and water. A scarcity discourse that emphasises further supply augmentation neglects the role played by skewed access to water in exacerbating scarcity for some and abundance for others. Through individual cases, the book illustrates the deprivation of marginal farmers versus the affluence of the large farmers. The increasing spread of private rights to water also reduces the possibility of institutionalising a system of collective ownership to water. The prevailing approach to scarcity also sidelines the issue of increasing demand for and consumption of water. The water crisis faced by paddy farmers is not viewed as an outcome of irrigation policies that have promoted the intensified cultivation of a water needy crop such as paddy. In the study area the availability of water through the canals led to a shift in the cropping pattern, with restrictions on the double cropping of paddy being done away with. The area has also witnessed an increase in the spread of water intensive coconut and plantain cultivation. Increased consumption of water is also an outcome of the enhanced abilities of resource rich farmers to invest in expensive energised lifting devices, which sanction unrestrained consumption of water.

Ownership and Use of Natural Resources

This section deals with the issue of distribution of existing water supplies. It discusses how the operationalisation of land and water rights in the area explains inequitable access to water and how it facilitates over-exploitation of a scarce resource.

When I commenced this study, the property dimension was furthest from my analytical framework. It is only while looking closely at changing land and water use patterns that I realized how important the concept of ownership was in understanding current resource management and use patterns. The existing rights regime over land and water in the area presented a significant stumbling block while envisaging of a possible sustainable and equitable mode of water resources management in the future. Benda-Beckmann and Benda-Beckmann have argued that any effort at redistribution of water among and within water use sectors as well as socio-political and eco hydrological regions would have to address the existing property rights regime over water (Benda-Beckmann and Benda-Beckmann 2003). Similarly, from the sustainability angle, the underlying rights regime is considered to pose a formidable challenge to efforts at integrated natural resource management (Klug 2002).

Amongst the institutional arrangements that mediate the human use of natural resources, the institution of property has been recognized as one of the most significant (Benda-Beckmann 2001, McCay and Acheson 1987, Berkes 1989, Hanna et al. 1996), determining not only who may use which resource and in what ways, but also shaping the incentives people have for investing in and sustaining the resource base over time (Meinzen-Dick and Pradhan 2001:10). Ownership over natural resources has been the subject of intense debate ever since Hardin wrote his essay on the Tragedy of the Commons. Attributing the over-exploitation of grazing lands to the absence of clear-cut property rights that governed access to the same, Hardin's theoretical propositions triggered off discussions on the relationship between property regimes and sustainable natural resource management. Hardin has been subsequently critiqued for depicting the tendency to free ride as inherent to human nature (Runge 1986), for equating open access with common property regimes and for propagating privatisation and/or state control as the only two ways by which the tragedy could be averted (Benda-Beckmann 2001, Meinzen-Dick and Bruns 2000, Sick 2007). A great deal of debate continues to surround the use and management of extensive, unpredictable and/or fluid common pool resources such as water systems, wildlife and fisheries (Sick 2007).

Initial discussion on property concepts was dominated by the 'big four' categories, viz. private, often individual ownership; state or public ownership; common or communal property and open access resources (Benda-Beckmann 2001), and their implications on sustainable resource management. Of late, it has come to be recognised that there exists no simple and linear relationship between particular property regimes and the way in which land and related resources are used and managed under each regime (Bromley 2002, Herring 2002). It has for instance been illustrated that state ownership does not ensure protection of the resource base, just as much as decisions of individual owners is no assurance for proper stewardship of land and related resources (Bromley 2002). To the contrary, in the context of private ownership regimes, Benda-Beckmann has argued that by granting free and unregulated access to scarce resources, the private property regime leads to the 'tragedy of individual ownership' (Benda-Beckmann 1992). The perception of resource degradation during the past decades in particular has brought into sharper focus the search for appropriate property regimes that could ensure sustainable resource management as well as greater justice and equity in the distribution of access to resources, particularly, forests, water and air (Benda-Beckmann et al. 1996). In fact a prominent focus of research on property rights and common pool resources has been on how to construct ecologically sustainable and socially just environmental regimes in the face of complex patterns of resource ownership and use, and complex social relations among multiple actors (Sneddon et al. 2002).

The following sub-sections discuss the increasing spread of the private property mode of resource ownership over land and water resources. They discuss the creation of the public and the private domains, and the manner in which such a categorisation leads to inequity and unsustainability in the management and use of water resources. They also discuss how the present formulation of rights to land and water, by neglecting the ecological properties of the resource, is inherently unsuitable to sustainable management of resources.

Extending the private

The increasing spread of the private property mode of resource ownership has led to the term property often being confused with something that is privately owned. The spread of European traditions in many parts of the developing world, had led to the private property model of land ownership being considered universal, with the notion of private individual ownership regarded as the 'apex of legal and economic evolution' (Benda-Beckmann et al. 2003)[12]. It is however important to note that even until a century ago, resources were governed by a wider range of ownership regimes. Studies into land tenure regimes across the world reveal that even in the not so distant past, land was subject to a variety of ownership and use regimes. In other words, private ownership and use of land was not the only institutional arrangement governing the use of land. The works of Netting, McKean, Jodha, Campbell and Godoy, reveal the diversity that existed in institutional arrangements that governed the use of land in locales as far away as the Swiss Alps, Nigeria, China, Japan, the arid regions of India, in the common field agricultural systems of England and the Andes.[13] In many cases, private individual or household property has been accompanied by corporate group rights in commonly owned resources (Netting 1993:158)[14]. Campbell and Godoy describe the common field agricultural systems of England where following harvest, private farmland was turned into communal pasture ground where all villagers were assured of grazing their animals, along with the collective rights of gathering peat, timber and firewood from common pastures and fallows (Campbell and Godoy 1986: 323). From far away south eastern Borneo, Vondal illustrates how a similar dual view of property classification evolved, that of private property in the dry, rice-growing season, and of unregulated, open-access common property in the rainy season, when water submerges the boundaries of privately owned and cultivated land. During this season, the submerged fields are used as common swamplands for fishing and duck farming (Vondal 1987: 247). Amongst the Bontoc Igorot of mountainous northern Luzon, forest lands were treated as common property with timber, firewood, basketry materials, medicinal plants, honey, pasture, mushrooms and game animals open to all village members, while wet rice cultivation and dry season potatoes, constituting three-fourths of the annual diet, was produced from irrigated, intensively worked terraces that was subject to a system of restricted inheritance (Prill-Brett 1986 in Netting 1993: 170). McKean while detailing on the community resource management systems of Japan offers an example of how individual households held rights to arable land, single villagers held shared rights to certain upland meadows and forests, and where groups of villages retained rights to other commons rather than sub dividing them into single village units (McKean 1996: 233).

These illustrations hold particular relevance in the case of farmland, which despite being 'inherently divisible' has not been managed on an entirely private basis in many parts of the world 'over remarkably long periods of time' (Campbell & Godoy 1986: 323). An important factor that emerges from these illustrations is that they exhibited a 'peculiar blend of private and communal endeavours' (ibid.) in the management of land and related resources (which includes forest land, grazing land and water). Communal property rights have often been dismissed as a sign of backwardness and economic inefficiency, as an obstacle to economic development and commercial production (Benda-Beckmann and Benda-Beckmann 2006), and the private mode of resource ownership portrayed as a solution to problems of resource degradation (Demsetz 1967, Hardin 1968). The individualisation of land tenure with registration of land titles was expected to encourage long-term investment in natural resource management and inhibit resource degradation. In addition to the sense of security conferred upon the landowner, it has often been assumed that the system of legal inheritance with the traditional father-son succession would ensure that the father would hand over to the son a viable enterprise (Bennett 1993: 132), thereby ensuring its sustainability. The conservation potential of private ownership has however been argued to be less operational when resources are subject to market mechanisms urging the owner to engage in exploitative use rather than caring for the ecological condition of the land (Bennett 1993) Widespread degradation of private land has been often been used as a point to vindicate the fact that private property does not always lead to the highest and the best use of the land (Bromley and Cernea 1989)[15].

The contested merits of the private property regime have not stalled its growing spread. Most human societies have moved toward what is becoming an increasingly similar landed property rights regime. As a result, complex, particularised local systems of property rights in land have been altered, transformed or replaced by simplified, more uniform sets of rules in a remarkably similar fashion across the world (Richards 2002).

The private and the public

Colonial interventions in land administration in many parts of the developing world have played a significant role in the present formulation of property rights over land and water. In India, prior to colonial interventions, control over land in most parts of the country was vested with regional chiefs or kings, while its actual use was mediated by a wide variety of tenurial agreements. The final user of the land possessed only usufruct rights to land, and was required to pay some kind of return to the owner or patron. When the British took over the administration of the country, they are reported to have found the existing system of land relations unsuitable to clear revenue assessments, as land parcels could not be identified with one particular claimant alone (Vani 2002). In an attempt to lay down clear principles to regulate the assessment and collection of revenue, colonial land policy aimed at institutionalising private, saleable property rights in land (ibid.)[16]. Individual agricultural producers and property holders were therefore critical to the British property system (Gilmartin 2003). The cadastral system of land survey, the system of laying out the land in rectilinear grids that was introduced in the country by the British, was intended to facilitate the easy determination of property ownership as well as the plotting of governmental jurisdictions (Bennett 1993: 133).

Along with the creation of private property rights to land, colonial land policies also led to vast tracts of uncultivated and forest lands being vested with the state (Iyengar and Shukla 2002). Hence, privatisation of property went hand in hand with the creation of state or public property (Herring 2002), largely over village forests and uncultivated lands. These lands fall in the 'public domain'[17], and are the only land portions that fall outside the ambit of private land ownership[18]. This has led to the creation of two broad property categories in the case of land, viz. the private and the public.

As with the case of land rights, existing water rights too have been moulded by European traditions. The colonial habit of making sharp distinctions between the public and the private realm, led to a situation where the public realm was ceded to the colonial government when it assumed sovereignty (Benda-Beckmann and Benda-Beckmann 1999). The history of water resource management in British India has been viewed as an irrigation history that established state monopoly over water resources (Whitcombe in Vani 2002). This overarching control was realised through the passing of statutes such as Irrigation Acts in different parts of the country for the regulation of water resources (Iyer 2003). The Northern India Canal and Drainage Act of 1873 was one such piece of legislation that declared the government's control over water flowing in rivers and streams, entitling the government to use and control such water for public purposes (Wescoat 2002, Vani 2002). Government control over flowing rivers has continued into the present. While it was only control over water resources and not ownership that was vested with the state, in practice the state behaved as though it was the sole owner, as manifest in the working of various state level irrigation acts (Upadhyay 2002).

Along with the creation of state property in water, the colonial land administration strategy also aimed at individualising water rights in tandem with the individualisation of land rights (ibid.). Hence, the distinction between the private and the public was made with regard to water too. Private waters are defined as waters located below, along or on privately owned land (Hodgson 2004: 49). This distinction was an important part of the common law tradition that influenced colonial land and water policy. In countries like India, the creation of the public and the private led to an asymmetry in the approach towards surface and groundwater (Iyer 2003: 103). While the law largely recognised only use rights and not ownership or proprietary rights over flowing surface waters, in the case of groundwater, ownership of land carried with it ownership of groundwater under it as well (ibid.). This has entitled land owners to sink a borehole or well on his land to intercept water percolating underneath his property even though the effect is to interfere with the supply of under groundwater to nearby springs (Hodgson 2004: 77) or to absolutely drain water from his neighbour's land (Torori et al. 1996).

Rights to groundwater have been conferred in fairly unconditional terms. The Indian Easements Act of 1882 for instance describes the right of the land owner with respect to groundwater as the 'right to appropriate water percolating in no defined channel through the strata beneath his land; and no action will lie against him for so doing, even if he thereby intercepts, abstracts or diverts water which would otherwise pass to or remain under the land of another' (quoted in Bhatia 1992). This differential treatment accorded to groundwater has come under heavy critique off late, largely due to the intensifying exploitation of groundwater (Bhatia 1992, Dubash 2007). It has also been argued to be particularly inappropriate to Indian conditions wherein small and contiguous land holdings share common aquifers (ibid.). In addition, deferred use by any one landowner only meant that another would capture it. The formulation of rights in such a manner implies that those with financial resources to sink wells are justified in appropriating a larger share of the water. Unconditional rights to the use of groundwater below one's landholding can also lead to depletion of water levels in water sources that share the same aquifer.

In addition to the public-private dichotomy over surface and groundwater systems, smaller water sources which are a mix of surface and groundwater have often been accorded a private property status. The ponds and the shallow pits (referred to as '*kuzhis*' in local parlance) in the study area are such examples. Being located on private land, they are considered as private property. Such a classification of small water bodies as private property is in line with ancient Roman or Islamic law. Ancient Roman law, which influenced the making of English Common Law, for instance, viewed seasonal water bodies as private (Hodgson 2004). Similarly, while Islamic law rejected the idea of private ownership over running water (Hodgson 2004: 48), water contained in receptacles or tanks, wells and springs developed by individuals on their own land was subject to private appropriation (Vincent 1990: 12–13). In the same vein, ponds and shallow pits in the study area, which were created by private enterprise in the past, continue to be viewed as private property. The present study shows how the private status accorded to these water sources exacerbates the inequity in access to water. In practice, owners of ponds and pits enjoy the same rights that owners of tubewells do. In addition, ponds help to extend the privatisation of common waters that flow down streams and canals. In so doing they contribute towards the shrinking of the commons.

The shrinking of the commons has been well illustrated in the case of land. In the Indian context, concerns have been voiced over the past two to three decades that public domain lands have shrunk by 26–63%, a considerable portion of which has ended up in private hands (Jodha 1986). The dominance of the private mode of land ownership has been at the cost of 'public' or 'non private' lands, which have been a highly valuable common property resource especially for the marginalized sections, helping to meet a substantial portion of their demands for fodder and fuel. This study shows that a similar phenomenon is underway in the case of water as well. It is not however easily discernible as in the case of land. The fluid nature of water makes it difficult to assess the extent of private appropriation.

The ecological dimension in property rights

While the existing public-private distinction leads to an inequitable distribution of access to water, it also results in unsustainable water use patterns. In the study area, the pigeonholing of different water sources into different property regimes leads to excessive exploitation of water from sources that are considered to be private. This exploitation is unmindful of the implications on water flows in adjoining water sources.

Of the three prominent theoretical categories assigned to natural resources (private, public and common pool resources), water has been defined as a common pool resource. Joint use and subtractive benefits constitute the defining feature of common pool resources,[19] and water exhibits both these features. The existing categorisation of water sources as public or private takes little consideration of the fluid nature of water, and of the basic hydrological interconnectedness of surface and groundwater systems. Unlike land and livestock, water moves freely across many different pieces of land, its quantity varying from season to season, even day to day (Anderson and Hill 1977). The flowing and transient nature of water, it being easily 'lost' as a result of evaporation or seepage into porous substances (Huggins 2000, Lam 1998) makes it difficult to subject it to clear-cut property definition. Despite these unique properties of water, formulation of rights to different water sources assumes that they exist in hydrological isolation. The existing public-private divide has therefore been viewed as a matter of 'hydrological nonsense' (Hodgson 2004).

The same is the case with land as well. Freyfogle has critiqued the practice of conceiving landowner rights in abstract terms, paying little attention to the natural features of the land parcel (Freyfogle 1999:31). The rights of the landowner are considered to be supreme as a result of which rights to land are not defined with respect to the natural features of the land (Freyfogle 1996). This has important implications as far as sustainable land use is concerned. The issue of paddy land conversion as well as the reorganization of rights to agricultural and forestland has been taken up later in the book to illustrate this point.

Of natural resources on the whole Klug argues that the formulation of legal rights remain rooted in the superiority of human entitlements, paying little attention to the needs of long term resource protection (Klug 2002). This he argues is largely due to our limited understanding of the ecology of the resource. This book argues that the formulation of land and water rights needs to be based on a finer understanding of the ecological properties of these resources. This is argued to be important if sustainable resource use is to be ensured.

Equity in Access to Water

Water rights play a critical role in defining who has access to water and who does not (Hodgson 2004). Water rights have been defined as a type of property right that aims, along with other water institutions and 'landed property rights', to assign access, use, liability and control over water from some persons and social groups relative to others (Wescoat 2002). Inherent uncertainty regarding its quantity and location coupled with demand for specific amounts of water at specific times and locations makes the issue of water rights a highly complex and contentious one (Meinzen-Dick and Pradhan 2001: 13).

Water rights are closely linked to rights to land as well as rights to the use of irrigation infrastructure. In the study area, this includes reservoirs and canal systems, ponds, energised tubewells and mechanised pumps. These play a critical role in ensuring access to water. Access to water may be defined as the availability of water in the 'right quantity and quality, at the right moment, and in the right place' (Van Koppen 1999: 5). When the infrastructure is made available through the state as in the case of state run irrigation agencies, then all farmers in the command area are entitled to receive a share of the water. But when individuals make investments in infrastructure, as in the case of the tubewell, then the ability to invest often determines access to the infrastructure and thereby to the water itself. Access to and use of water is therefore embedded in unequal social, economic and political contexts (Mehta 2000).

The previous sections have illustrated how the existing property categorizations determine access to water. In addition to property relations, the access to and use of water is mediated by the social and economic status of users as well. Crow and Sultana outline four different modes of access to water: a) ownership of land and access to a pump for accessing surface and groundwater, b) market access to water through purchase, c) common property access to sources such as rivers, ponds, public tanks through communal rights of access and d) state provision of water which gives access to national or local government projects. They also emphasize that access to water in all these four modes is differentiated by material and social conditions of access, with the rich often benefiting from preferential conditions of access to multiple sources of water (Crow and Sultana 2002).

Of these above mentioned factors, ownership of land, access to irrigation sources that are designated as 'private', along with access to energised lifting are the three factors that lead to skewed access to water in the study area. I argue that the existing system of water rights that are tied to private rights to land gives the landed increased access to water sources such as ponds and wells. This is further aggravated by the fact that the well to do farmers are able to invest more in costly energised tubewells and in energised pump sets, giving them enhanced access to privately owned groundwater reserves and commonly owned resources such as streams.

The implications of such a trend can get severe when the existing supply of water is decreasing. This means those with access to privately owned water sources and the ability to invest in extractive technologies get to eat more from a shrinking pie. This strengthens the argument that private interests need to be curbed in the light of common interest (Benda-Beckmann F. von 1992). This study argues that the existing skewed nature of access to water requires a reconsideration of the existing public-private classification over water sources, keeping in mind in particular, the extractive potential of modern energised pumping devices.

The Research Site

The study focused on one of the prominent paddy pockets in the southeastern part of Palakkad district, viz. the Kollengode and Elavenchery panchayats of the Chittur taluk in Palakkad district.[20] The Kollengode and Elavenchery panchayats cover an area of 50.65 and 32.15 sq. km respectively (GOK 2007a). Within these two panchayats, the Varayiri watershed covering an area of 1450 ha was chosen for detailed field study. The Varayiri stream is a tributary to the Gayatri River, which is located in the Bharathapuzha river basin (see Map 1.1). The Bharathapuzha river basin is considered to be one of the most degraded river basins in the state, having experienced deforestation and land use changes of a significant nature. The following sections will introduce significant features of the ecological and social landscape in which the study was conducted.

Climatic Peculiarities of the Palakkad Gap

The presence of the Western Ghats at the eastern boundary of the state of Kerala exercises the most decisive influence on the climate and topography of the state. The Western Ghat mountain ranges run almost parallel to the west coast of the country starting in the north from near the Tapti River in Gujarat and ending in the south near Kanyakumari in Tamil Nadu (S. C. Nair 1991). Amongst the south Indian states, the state of Kerala can be considered as one of those, more or less exclusively governed by a tropical monsoonal climate.

The presence of the Ghats (rising up to an average elevation of 900–1500 metres above sea level) creates an orographic barrier to the moisture laden winds that blow eastwards from the Arabian sea, precipitating a significant amount of downpour on the western side of the Ghats, during the period of the south-west monsoon (June–September) (ibid.). Being confined to the western side of the Western Ghat ranges, most parts of the state receive annual rainfall of a very high order (2500–3000 mm a year).

The region enclosed within the administrative boundaries of the Palakkad district and within the natural boundaries of the Bharathapuzha river basin stands out as an exception, owing to the presence of the Palakkad Gap. The Palakkad Gap is a unique formation, where the mighty mountains recede for a short distance[21] (see Map 1.2). The absence of the Ghats for about 45 kms along the eastern boundary of the district has precipitated unique climatic conditions in the Palakkad plains[22], especially in terms of wind and

Map 1.1 India, Kerala, Palakkad and the Study Area

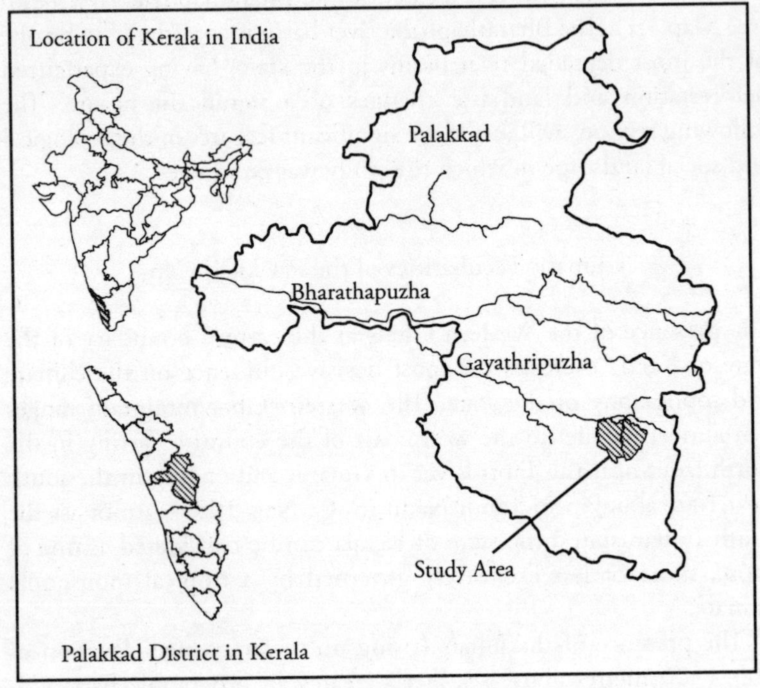

rainfall patterns, which distinguishes it from the rest of the state (S. C. Nair 2004: 7).

The average rainfall of Palakkad is significantly lower than the Kerala average. During the years 1998–2004, while the state of Kerala recorded an average annual rainfall of 2718.66 mm, the average annual rainfall for Palakkad was 2016.85 mm (GOK 2007a). During certain years the actual rainfall received was less than this average, falling to 1831 mm in 2000 and 1728 mm in 2003 (ibid.). Even this annual average for the district of Palakkad cannot be taken as an approximation for the rainfall figures in the study area (i.e. the panchayats of Kollengode and Elavenchery). This is due to the wide variations in rainfall between the western and eastern parts of the district.[23] The average annual rainfall recorded at the Kollengode

Map 1.2 Location of Palakkad Gap in Kerala

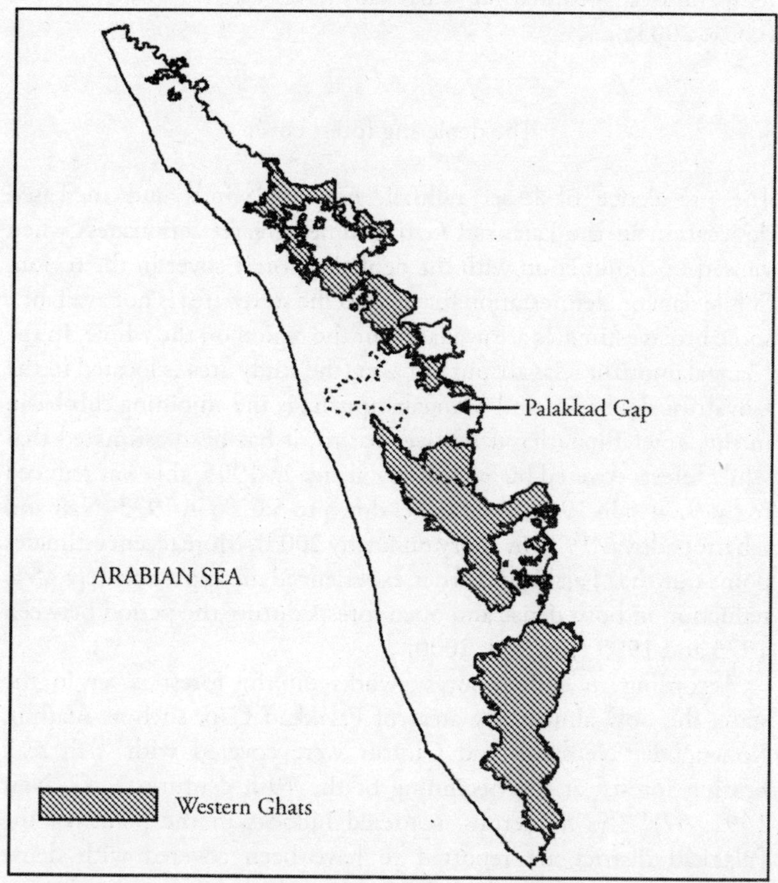

Source: S. C. Nair 1991

station during the above-mentioned period was only 1580.27 mm, far below the district and state average.

Apart from receiving a lesser amount of rainfall, the area is also subject to 'continuous dry high velocity winds' (ibid.) funnelled in through the Gap from the neighbouring Coimbatore plains in Tamil Nadu. As a result, during summers, the district of Palakkad, particularly those portions located in the Gap record higher temperatures than most other parts of the state. Maximum summer temperatures recorded at Palakkad have often crossed 40 °C (GOK 2003a).

The depleting forest cover

The prevalence of lesser rainfall, hot dry winds and increased desiccation in the Palakkad Gap assumes greater seriousness when viewed in conjunction with the depleting forest cover in the region. While data on deforestation for the specific study area is not available, some broad estimates are available for the region on the whole. In the Mangalampuzha-Gayatripuzha basin (the study area is located in the Gayatripuzha basin, and Mangalampuzha is the adjoining sub-basin in the larger Bharathapuzha river basin), it has been estimated that while forests covered 60% of the total area in 1905, this was reduced to the 9.74% in 1965, and further down to 5.62% in 1973 (Nair and Chattopadhyay 1985 in Sooryamoorthy 2003). More recent estimates point out that Palakkad district experienced an approximately 45% reduction in both dense and open forests during the period between 1973 and 1995 (Jha et al. 2000).

According to contemporary works on the forest cover in the state, the now almost dry areas of Palakkad Gap, such as Alathur, Kollengode, Nenmara and Chittur were covered with 'rich teak bearing forests' at the beginning of the 19th century (S. C. Nair 1991: 57). The numerous, scattered hillocks in the plains of the Palakkad district are reported to have been covered with dense vegetation as recently as 1971, prior to the nationalization of private forests.[24] Names of local places in the study area, like Tekkinkadu (meaning teak forests), Vattekad and Karekad (*kadu* meaning forest) in the area indicate the presence of forests in the past. The fact that small hillocks like the Mookarshan hill and Cheerani hill in the Varayiri watershed were well covered with trees in the memory of old farmers is testimony to the presence of forests even in the recent past.

Descriptions of the forest cover are found in Francis Buchanan's account of the region when he surveyed parts of southern India in

1800–01,[25] indicating the vegetation pattern that once existed here. While describing his journey forward from Kollengode to Palghat, Buchanan wrote,

> I went a long stage to Pali-ghat. The country through which I passed is the most beautiful that I have ever seen. It resembles the finest parts of Bengal; but its trees are loftier, and its palms more numerous. Not only in the mountains, even the landscape lower down seems to have been well covered with trees. 'The woods through which we passed to-day are very fine; Teak and other forest trees are now fast springing up among the Banyan . . . and Palmira trees. . , by which the houses of the natives have formerly been shaded; and this part of the country will soon be longer distinguishable from the surrounding forests' (Buchanan 1870: 49–50).

In another instance he wrote, 'The hills toward the south are covered with trees to the summit' (p. 76). The hills to the south that Buchanan referred to are the mountains that form the southern rim of the Palakkad Gap. This mountain range is also known as the Tenmala, which when translated mean the mountains that border the southern part of the region. The Tenmala is the most distinguishing feature in the region around Kollengode (see photo 1.1).

A century later, C.A. Innes, the Settlement Officer of the erstwhile Malabar district wrote the Malabar District Gazetteer. In this work he mentions how elephants in the Kollengode forests (i.e. the forests of the Tenmala) were plenty enough to comprise a valuable source of income to the landowners (Innes 1908).[26]

The setting up of *swarupams* or small kingdoms such as the Kollengode *swarupam* has been traced to the 17th century when cultivation was expanding into the Western Ghats, leading to forest land being cleared for agriculture (Ganesh 1991: 305). Coupled with the loss of forests in the upper catchments of almost all the tributaries of the Bharathapuzha River, forests in the plains disappeared with the expansion of cultivation, particularly paddy, as well as the nationalisation of private forests. This 'killed off all the smaller watersheds' (S. C. Nair 2004) like the Varayiri in the basin, the newly exposed rocky terrain heating up and adding to the desiccation of the Palakkad Gap.

Photo 1.1 The Tenmala

Two hundred years after Buchanan wrote his account of this region, it is hard to believe that 'lofty forests' once grew here. The Tenmala that was reportedly covered with trees to the summit stands bare today. From a situation when the tree cover in the country was considered dense enough such that it 'will soon be no longer distinguishable from the surrounding forests', the only trees that remain are the palymra, which too are fast decreasing in number.

The dry hot winds that blow across the Gap further desiccates the land that has lost most of its tree cover, especially so as the adjacent Coimbatore plains are even more severely desertified (S. C. Nair 2004).

> In the Palghat taluk especially during February, March and April a hot wind rushes in from the burning plains of Coimbatore, and dries up every green thing for miles around (Kareem 1976: 12).

The irrigation scenario in Kollengode and Elavenchery

A relatively lower amount of rainfall, warm temperatures and increased dryness due to the hot winds that blow over the region make water economy crucial in the region. Seasonally concentrated rainfall, in addition to the relatively lower amount of rainfall in the eastern, drier parts of Chittur taluk gave rise to early trials with irrigation.[27] While rainfall along with minor diversions from rivers and streams met the agricultural water requirements in most parts of Kerala, this was not the case with the eastern parts of Palakkad district where kulams/eris (ponds/tanks) have been used for irrigation purposes.

Though ponds are found in many other parts of Kerala, they were largely used for bathing, watering cattle and by washer folk, with irrigation not figuring as a prominent use (Unni 1965 in Mencher 1998: 45). The common function that links the kulams of Palakkad with the *eris* of Tamil Nadu, Andhra Pradesh or Karnataka, the *ahars* of south Bihar or the *khadins* of Rajashtan is that of harvesting run off water by building an embankment across the line of drainage so as to hold back surface run off. An important function of these systems is to sustain agriculture by supplementing and enhancing the predictability of water supply. One of the main differences between the kulams in Palakkad and the eris of Tamil Nadu or Karnataka (commonly referred to as tanks) is that the former occupy a smaller surface area, having smaller catchments and irrigating smaller commands. It is for this reason that the word pond has been used instead of tanks in this book, while referring to the kulams. While some of the larger ponds in the study area are referred to as eris, they occupy a much smaller surface area than the eris of Karnataka or Tamil Nadu.

The similarity in construction and in the technology of sluices found between the kulams of Palakkad and the eris of other south Indian states, which is not observed in the case of kulams found in other parts of Kerala, leads to the supposition that it is the Tamilian influence, facilitated through the Palakkad Gap that led to the emergence of this irrigation technology in the eastern part of the Palakkad district. Shencottah, in south Kerala, located in yet another gap in the mountains, is an area where ponds/tanks employing a similar technology are found. As noted by Kuriyan,

> These regions (Shencottah, Chittur and Palghat) happen to be the very parts of Kerala where there is maximum possibility of contact with Tamilnad. . . . Apparently therefore there exists a considerable amount of correlation between agricultural practices like the methods of ploughing, irrigation, etc. and linguistic and cultural distributions (Kuriyan 1942 in Mencher 1998:45).

Mention of the ponds is found in Francis Buchanan's account as well. Buchanan made note of the ponds as –

> Here the rain, without any assistance from art, is able to bring one crop of rice to maturity; and in a few places the natives have constructed small reservoirs, which enable them to have a second crop' (Buchanan 1870: 50).

And in another instance,

> . . . in the lower grounds a second crop of rice may be depended on, wherever small reservoirs have been constructed to give a few weeks supply toward the ripening of the corn after the rainy season has abated (ibid., 69).

Innes too has made a mention of ponds when he wrote The Malabar Gazetteer in 1905.[28]

Palakkad's distinction as one of the important rice growing areas of Kerala dates back to the pre modern irrigation era. As early as 1904–05, the *taluk* of Palakkad[29] (the administrative boundaries of which do not coincide with the present day Palakkad taluk), contained the greatest proportion of wetlands used for paddy cultivation (Innes 1908: 208). Since the 1950s, the advent of modern irrigation centred on the construction of reservoirs and canals, has placed Palakkad upfront on the irrigation map of the state. Palakkad is the district with the highest number of irrigation projects in the state (a total of seven major and medium irrigation projects)[30], all of which are aimed at the expansion and intensification of land under paddy cultivation.

All of these dams have been constructed across various tributaries of the Bharathapuzha River. Of the seven large and medium scale irrigation projects in the district, five are located in the Chittur taluk itself. The panchayats of Kollengode and Elavenchery fall in the command area of the Gayatripuzha Irrigation Project.

Though the case of Palakkad has been portrayed as a successful example of canal irrigation in the state, almost all the irrigation systems perform far below their intended capacity (Vishwanathan 2002). Deforestation in the upper catchments of all the rivers that have been dammed has led to high rates of siltation, significantly reducing the storage capacity of the reservoirs (Kumar 2005). Nair reports that routinely after the monsoon, there is not enough water to be distributed in the command areas of the dams. Dams that were originally constructed for irrigation are now more vital as drinking water sources (S. C. Nair 2004). The irrigation scene in the Chittur taluk has been further complicated with the inter-state conflicts over the waters of the Aliyar River (tributary to the Bharathapuzha which originates in Tamil Nadu) governed by the Parambikulam-Aliyar agreement on water sharing between the governments of Kerala and Tamil Nadu.

Despite the existence of an irrigation history based on ponds prior to the introduction of modern canal irrigation in the district, the former have received scant attention. The database regarding the ponds is scant and not very reliable. The District Gazetteer for Palakkad for instance refers to them as follows:

> As for tanks, there are large numbers of them all over the district which are used for drinking, bathing and irrigation purposes. According to a recent estimate there are 99 such tanks in the district' (Kareem 1976).

Ninety-nine ponds/tanks for the entire district is a gross underestimate, there being 363 and 268 such ponds in the Kollengode and Elavenchery panchayats alone (Kerala Land Use Board 2001). Another estimate of ponds in the state on the whole, has indicated the greater prevalence of relatively larger ponds in Palakkad (CWRDM n.d.). Of the total number of ponds (910) above 0.5 hectares in area in the state, roughly 66% (600) are located in Palakkad district (ibid.).

Canals, ponds, wells and streams

Similar to most of the southern and eastern parts of Chittur taluk, hydrologically interconnected irrigation systems, viz. the canals, ponds, streams and wells provide water for agriculture in the watershed (see Map 1.3). The drainage into the Varayiri stream is intercepted by 150 odd ponds of varying sizes (the average area ranging between 0.4 and 1.2 hectares), which were the primary sources of irrigation in the area prior to the introduction of canals in the late 1960s.

Modern canal irrigation was introduced into the area through the Gayatri Irrigation Project in the late 1960s. The left bank canal of the Gayatri Irrigation Project traverses through the southern ridge of the watershed (see Map 1.3). The Varayiri watershed falls in the command area of the four distributary canals (viz. the Kollengode, the Payilur, the Peringotukavu and the Karinkulam distributaries) that are located in the second reach of the left bank canal. Apart from ponds and canals, wells (both shallow and deep), and streams comprise critical sources of irrigation. The spread of energised pumping from wells, streams and ponds has also enhanced irrigation possibilities in a significant manner.

In addition to the frequent reporting of water scarcity from the Chittur taluk, the co-existence of modern and traditional irrigation made this region an interesting case for irrigation studies. The commonly cited reason for water scarcity was the non-availability or delayed availability of irrigation water through the canals. The neglect of the traditional irrigation structures (i.e. the ponds), an inadequate appreciation of the ecological and agricultural functions of the ponds, changing land use and cropping patterns and changes precipitated by the introduction of the canal irrigation systems seemed to be important starting points for a study on recurrent water scarcity. Since canal irrigation was introduced to the Kollengode and Elavenchery panchayats only during the late 1960s[31], it enabled a comparative analysis of the functioning of ponds and canals, as well

MAP 1.3 Canals, Ponds and Streams in the Varayiri Watershed

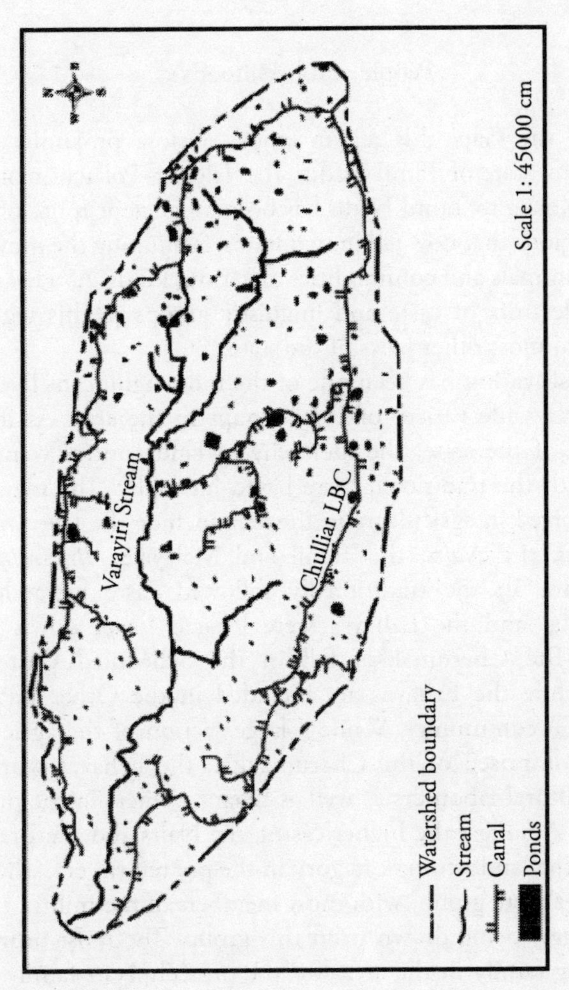

Source: Modified from Panchayat land use maps, State Land Use Board 2001.

as the changes precipitated by the introduction of the canal systems. It was felt that studying the transition from pond to canal based irrigation would provide insights into the changes that have come about to water resource management in the area.

People and livelihoods

Located in the Gap, this region enjoys a close proximity to the neighbouring state of Tamil Nadu. The Trichur–Pollachi main road that links Kerala to Tamil Nadu has been an ancient route of traffic between regions that now fall in two states, facilitating the movement of people, animals and commodities of various kinds. As a result, one finds a wider mix of caste and linguistic groups in this region, as compared to most other parts of the state.[32]

Paddy cultivation has been one of the most significant livelihood options for a wide variety of caste groups in the area. As in most other parts of the state, the hierarchy of land control went hand in hand with the traditional caste based hierarchy. The main caste groups involved in agriculture in the area include the *Cherumakkal*, the *Ezhavas*, the *Nairs*, the Tamil and Malayalee *Brahmins*, and the Muslims. In the traditionally followed caste hierarchy, the Cherumakkal and the Ezhavas were located lower down in the hierarchy. The Cherumakkal fall in the Scheduled Caste (SC) category, while the Ezhavas are included in the Other Backward Class (OBC) community. While a large section of the agricultural labourers comprised of the Cherumakkal, the Ezhavas worked as both agricultural labourers as well as tenant farmers in the pre land reform era. Amongst the higher castes, the Nairs and the Brahmins comprised the landowning category in the pre reform era. The Nairs are an upper caste group, with most members of the militia and the land managers being drawn from this group. The most prominent land owning family in the area was the princely Nair family of the Kollengode *Kovilakam* also known as the *Venginad Kovilakam*. The Brahmins of the area belong to the Tamilian Brahmin community (locally known as *Pattar*) and to the Malayalee Brahmin community referred to as the *Namboodiris*. Amongst the Brahmins, the Tamil

Brahmins dominate in numbers, living in clusters known as *taras* that are dispersed across the region. Some of the larger tenants of the area also belonged to the Nair community, who sub let their holdings to smaller tenants who mostly belonged to the Ezhava community. Muslims had also made their presence felt in the region in the pre reform era, but mostly as agricultural labourers or small tenants.

With the implementation of the land reforms, hitherto tenants acquired landowning rights. Following the land reforms, the earlier caste-based hierarchy no longer paralleled the hierarchy of land control. The Ezhavas therefore comprise a significant proportion of the land owning community in the area today. While the Nairs and Brahmins retained their land holding status (despite having had to surrender land in excess of the fixed ceiling), the spread of modern education has led to a situation wherein most of them have moved out of active farming.

A significant outcome of the implementation of land reforms in 1970 has been the fragmentation of land holdings. In the Kollengode Block Panchayat[33] for instance, approximately 80% of the individual operational holdings were below 0.5 ha (1.25 acres) covering 10.41% of the total area (GOK 2001a). The remaining 20% of the holdings are located on 80% of the land, with only 8.72% of the holdings covering an area of more than 2 ha.

Apart from farming, there existed a wide variety of caste based livelihood options. These included the *Koravar* community who were bamboo artisans, the *Kumbaran* caste group who were potters, the *Asaris* who were traditional carpenters, the *Kollan* caste who were blacksmiths (Kollengode being famous for the high quality of knives made by this group of artisans), the *Chettiars*, the *Pandaranmaar* and the weaving community. Some of these caste groups such as the Chettiars, the Pandaranmaar and the weavers migrated into the region through the Gap. In addition, toddy tapping was an exclusively Ezhava occupation. Similarly, the Muslim community of the area was traditionally engaged in the fish business. The settlement pattern of the area was also caste based, with each caste group residing in more or less well defined clusters. The Brahmins residing in *gramams*, the weaving community in *pavadis* and the Nairs in taras. The agricultural labourers mostly live in clusters referred to as *kudiyirippu*.

Caste based livelihood options are slowly changing with the advent of modern education and with the increasing job opportunities available in the nearby towns and cities of Pollachi and Coimbatore. While the Nairs and the Brahmins moved away from agriculture about three to four decades ago, the present day younger generation of all the other communities are gradually moving away from their traditional livelihood occupations.

Method of Study

This study developed as a qualitative survey of changing land and water management practices in the Varayiri watershed situated in the panchayats of Kollengode and Elavenchery in the Chittur taluk of Palakkad district in Kerala (see Map 1.4). Preliminary field visits to different parts of Chittur taluk led to the selection of this area for detailed study.

As mentioned earlier, not all parts of Palakkad exhibit a concentration of ponds. They are found in larger numbers in the southeastern taluk of Chittoor. The panchayats of Mudalamada, Kollengode, Elavenchery and Peruvamba exhibit a high concentration of ponds. The Kollengode panchayat contains the highest number of ponds in the district, 363 in all. The absence of any detailed database on the ponds implied that I had to rely on direct explorations.

Most of the existing sources of irrigation are hydrologically interconnected, with water stored in one source recharging other sources. The water supplied through the canal network seeps into the streams below, similarly water stored in ponds raises water levels in near-by wells. While observing water management practices in the Kollengode and Elavenchery Panchayats, one could see that the canals, ponds, bore wells, shallow pits in the valley bottoms, wells and the stream all formed parts of a larger design. The question in my mind was – 'How do these different water sources relate to one another and to paddy cultivation in the area? I can see a canal passing by here, a few ponds there, shallow pits in another place, but what

Map 1.4 Varayiri Watershed within the Kollengode and Elavenchery Panchayats

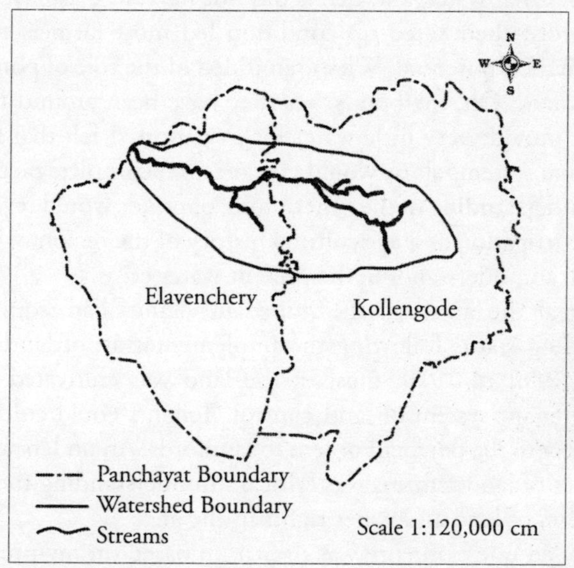

Source: Modified from State Land Use Board 2001

does this tell me with respect to water resources management?' To have a better understanding of the functioning of these different water sources, I felt that it was important to locate my field enquiries within the boundaries of a watershed. Fieldwork was therefore conducted within the natural boundaries of a watershed and not within the administrative boundaries of a panchayat. The area draining into the Varayiri stream was selected for detailed study.

My understanding of the land and people of Kollengode and adjoining areas began with my interaction with the farmers of Nenmeni in the Kollengode Panchayat, where I was staying. Fieldwork was conducted in the 2001–2002 period. When I explained my area of interest to the farmers of the area, they would immediately begin discussing the problems they encountered with water supply, primarily with the supply of water through the canals. It soon became clear to me that the farmers were more inclined to talk of the canals than the ponds. The reasons were clear. For one, ponds irrigated a very small area as compared to canals. Secondly, ponds were today being filled with canal water, and hence did not have an existence of their own. Thirdly, their silted up condition led most farmers to under emphasise their potential. When reminded of the role of ponds, they would remark 'Oh, the ponds, yes they have been around for some time, but provide very little water for irrigation'. I felt that focusing on the canal systems alone would give me an incomplete picture, and that an understanding of the functioning of ponds would reveal more about the irrigation and agricultural history of the region, which was important in understanding the present water crisis.

Almost all the farmers in the immediate vicinity had acquired their land holding status following the implementation of land reforms in 1970. Prior to 1970, most of the land was cultivated under a landlord-tenant system of land control. Tenants could cultivate the land subject to the payment of rent to landlords. An understanding of the impact of land reforms was critical in understanding the current formulation of land and water rights in the area.

A detailed questionnaire was drawn up based on my preliminary interactions with farmers of the area. During the detailed farmer interviews that were to follow, issues related to changing land and water management practices were discussed in detail. Discussions on land management focused on the changes in the cropping pattern, the intensification of paddy, cropping patterns in pond irrigated areas in the pre canal era, canal induced changes in the cropping pattern and the agricultural calendar. Discussions on water management focused on the present functioning of canals, the functioning of ponds in the pre canal and post canal era, the spread of tubewells, existing system of water distribution in the canal and pond systems and the nature of conflicts over water rights and water distribution. The process of implementation of land reforms and the distribution of land and water rights in the pre land reform era, were discussed in detail. Apart from the farmer interviews, discussions were also conducted with some of the aged farmers of the area who were able to give details of the changing rights regime.

Given the skewed nature of land holdings in the area, farmers from all landholding categories were interviewed. Farmers to be interviewed were selected from the list of paddy farmers available with *padashekhara samities* in the area. *Padashekharam* refers to a cluster of paddy fields and padashekhara samities are registered bodies of paddy farmers cultivating an area of 50–100 hectares on an average. Each padashekara committee was registered with the *Krishi Bhavan* (the Agriculture Office for each panchayat). Sixteen padashekhara samities were found to function in the Varayiri watershed, which were registered with the Krishi Bhavan of either the Kollengode or Elavenchery panchayats. A sample of 9 farmers were drawn from each of these padashekhara samities with three farmers each drawn from three landholding categories, viz. farmers with holdings less than one acre (0.4 ha), with holdings between one and five acres (0.4 ha – 2 ha), and with holdings above five acres (2 ha). The existing classification in the state classifies farmers as large, medium and small as those with holdings above 2 ha, between 1–2 ha, and below 1 ha respectively. For this study however, I have classified small and marginal farmers as those with landholdings below (0.4 ha), and not 1 ha as a majority of the small farmers in the study area have holdings below 0.4 ha. In Elavenchery panchayat for instance, 72% of the farmers owned land less than 0.4 ha (Elavenchery Panchayat, 1996). There seemed to be a significant difference between farmers with holdings below 0.4 ha, and those with holdings between 0.4 ha and 1 ha, as far as access to land and water were concerned. Grouping them together into a single category would therefore not bring out sharply the existing inequities.

Farmers belonging to these three landholding categories were therefore selected from all the sixteen padashekhara samities whose lands drain into the Varayiri stream. It needs to be noted here that the boundaries of the padashekhara samities do not conform to watershed boundaries. However, except for 2–3 padashekhara samities, most of them were largely confined within the Varayiri watershed. While the interviews conducted with the 144 farmers selected in such a fashion revealed many details about the ongoing land and water management practices and the changes therein, the landholding based selection gave further details on differentiated access to land and water. This was further substantiated with the collection of detailed farmer histories of select farmers from each landholding category. These interviews were supplemented with a full time direct observation of water management and water distribution practices in the different irrigation systems in use in the area, during the period of fieldwork. This involved in particular a detailed observation of water management practices in the Gayatri Irrigation Project.

Structure of the Book

This book consists of seven chapters. Following the introductory chapter which introduces the theoretical concerns of the study along with a description of the socio-ecological context in which the present study has been undertaken, the first part of the book (chapters 2 and 3) discusses changing ecology, hydrology and water control in the study area as an outcome of modern irrigation and agriculture policies. Chapter 2 presents the logic behind the spread of modern irrigation and the intensification of paddy cultivation in the state. It then discusses the impact of these policies on land and water use patterns in the study area. Chapter 3 discusses the impact of infrastructure oriented irrigation polices on the sustainability of water resources. It shows how the neglect of the sustainability dimension in the management of each of these irrigation sources has negative implications on the long term sustainability of local water resources, which therefore aggravates the problem of water scarcity, manifest in the declining storage of water in the reservoirs and ponds, and declining flows in the streams. It also discusses how the implementation of governance reforms through decentralisation policies has made little change in the existing approach towards water resources management.

The following three chapters (Chapters 4, 5 & 6) discuss the impact of various social dynamics in shaping access to water. A major part of the field data is presented in these chapters, with each chapter showing different aspects of land and water rights, and their implications on the distribution of water. As they evolve, Chapter 4 discusses the emergence of permanent rights to land and water and a consolidation of their private nature, Chapter 5 elaborates on the inequities in access to canal water, and finally Chapter 6 discusses the growing spread of private rights to water.

Chapter 4 presents historical details on changing agrarian relations in Kerala, in order to illustrate the changes in rights to land and water. These details also provide the context for a part of the analysis on public and private rights to water in Chapter 6. It discusses the inter related nature of land and water rights in the study area, and the changes brought about through the implementation of land reforms in the area in 1970. It discusses how the conferring of private land titles created inequities in access to water. Chapter 5 looks closely at the existing water distribution practices in the Chulliar canal system that is a part of the Gayatripuzha Irrigation Project. It illustrates how location in the irrigation system along with the operation of unequal power relations shape access to water delivered through the canals. Chapter 6 is the last of the three chapters that deal with factors that contribute to the inequitable distribution of water in the study area. This chapter reviews the existing property classifications over water (the public-private classification), and illustrates the manner in which the property-technology interface shapes access to water in the area. It also shows how this interface has resulted in private rights gradually encroaching upon commonly owned water resources.

Chapter 7 concludes the book.

Notes

1 Puzha in Malayalam refers to river.

2 The other being Kuttanad, the low-lying paddy tract in the southern district of Alapuzha.

3 As of 2004–05, roughly 38% of the total area under paddy in the state was located in the district of Palakkad, contributing 39% (the highest share) towards the total production of paddy in the state (GOK 2007a).

4 Some of the typical news reports on this issue during the summer months read as follows: '*Vellam ilya; vaidyuthiyum; karshika mekhalekku venal pediswapnam' (In Malayalam).* (No water; no electricity; the summer is a nightmare for the agricultural sector') (Mathrubhoomi February 13th, 2002). Reports such as these are very common during the summer

months. Another such report, *'Kathir Urakkum Mumbe Vellam Nirthi; Malampuzha Ayacuttil Krishi Unakkam'* (The water (supply) was discontinued before the formation of the pannicle; crop dries in the command area of the Malampuzha Project) (Mathrubhoomi, March, 12th, 2002).

5 In most parts of the state two crops of paddy are raised. The first crop (autumn crop) known as the Virippu is cultivated between the months of April/May and August/September. Most of the growing period of the first crop coincides with the south west monsoon. This is followed by the winter crop known as Mundakkan, also known as the second crop, cultivated between the months of October–January. In the low lying coastal areas, a third crop known as the Punja is cultivated between January–April. In the midland zone, the Punja crop is taken only in those areas with an assured water supply.

6 The lowest tier in the three-tier structure of governance constituted at the village level.

7 During the period between 1995–2002, farmers from Kolumbu, Karinkulam, Vattekad and Panangattiri in the Elavenchery panchayat lost their second crop of paddy on four occasions. The area affected by crop loss covered approximately 400 hectares. On all these occasions, the Krishi Bhavan (the Agricultural Office) has paid compensation to the affected farmers (Data collected from the Elavenchery Krishi Bhavan).

8 Taluk is a revenue division within a district.

9 Many of the downstream functions of rivers such as delivering nutrients to the sea, absorbing floodwaters, protecting wetlands, providing habitats which maintain the rich diversity of aquatic life and maintaining salt and sediment balances (Postel 1999: 119) have been disrupted with the construction of dams and reservoirs upstream. The significant reduction in the total amount of fresh water reaching the seas across the world is an indication of the severe downstream impact of dams and diversions (Postel 1999:71). The recognition of the importance of natural flow regimes to riverine health has led river restoration to become an expanding area of investment by public water management bodies in many developed countries (D'Souza R 2003).

10 The allocation for minor irrigation was only 17% of the total allocation for irrigation in the early years, but grew to 24% by the Ninth Plan (GOK 2002a).

11 In Kerala, the percentage outlay for the irrigation sector was 6.38% in the Ninth Plan, which declined to 3.88% in the Tenth Plan (GOK 2006d).

12 Freyfogle argues that nineteenth century western historians believed that shared ownership was the primitive form of property, and individual ownership the more advanced form (Freyfogle 1996).

13 Much of this work emerged in response to the debate centered around Hardin's theory of the Tragedy of the Commons, to highlight that common property arrangements were not always 'tragic'. Their work on customary forms of land ownership also reveal that property arrangements have not always been bimodal, but have existed along a continuum (Bromley 89: 875).

14 Netting argues that 'with increasing scarcity, value and permanence of agricultural resources, rights are progressively delimited from the kin or local group to the lineage, from the lineage to the family, and within the family to single heirs' (Netting 1974: 42). While he observes that rules of ownership, exchange and inheritance tend to be increasingly detailed when such intensely used resources have a higher subsistence or market value, it was only intensively cultivated land that was subject to private property arrangements, with common pool resources such as grazing grounds, surface water, and forests being held as common property (ibid., 173).

15 Bromley and Cernea cite the example of large segments of the best agricultural land in Latin America being devoted to cattle ranching, while food crops are grown on poorer lands (Bromley and Cernea 1989).

16 The essence of private property rights to land, that is the right to transfer by sale or mortgage, had to be recognized by law, so that the state could enforce revenue collection by attaching and public sale the land of cultivators or owners (Vani 2002).

17 Bromley defines the public domain as 'all land not held in private ownership (freehold or fee simple) by someone. The public domain includes land administered by national/state/provincial governments, it includes land administered by villages as true common property, and it includes land that is managed by no one and hence properly called open access' (Bromley 1991: 108).

18 The use of lands falling in the public domain is regulated by the Village Panchayat, Revenue and Forest departments (Iyengar and Shukla 2002: 310).

19 Fish, wildlife, forests, grazing lands, irrigation and groundwater have been classified as common pool resources (Chopra et al. 1990, Berkes 1996, Wade 1988), which are characterized by joint use and subtractive benefits, and is potentially subject to crowding, depletion and degradation (Wade 1988).

20 The panchayat, block and the district represent the three levels in the three-tier system of governance followed in the country. The panchayat is constituted at the village level. The block is a sub division of the district and is therefore an intermediary level.

21 The discontinuity, which marks the Gap, gives rise to an abrupt fall in the mountain ranges from an average height of 1500 metres above sea level to a floor height of less than 100 meters above sea level within the Gap. This discontinuity is attributed to a likely tectonic origin through which a river could have flowed through ancient times (S. C. Nair 1991). 'Here by whatever great natural agency the break occurred, the mountains appear thrown back and heaped up as if some overwhelming deluge had burst through sweeping them to left and right' (Logan 1887).

22 It is the climate of the lowland areas of the district, which can be broadly referred to as the Palakkad plains, which are the most affected by the Gap formation. The highland areas are confined to the Attappady and Nelliampathy hills in the Western Ghats, essentially located on either side of the Palakkad Gap.

23 During the fifty-year period between 1901–1950, while the average annual rainfall recorded at the Mannarkad rain gauge centre in the western part of the district was 2890 mm, it was only 2115.2 mm at Palakkad and 1794.1 mm at Chittur in the eastern part of the district (Census of India 1961). The variation is significant also in terms of the average number of rainy days in a year, varying from 122 in Mannarghat (located in the western part of the district) to 93 in Chittur (Kareem 1976: 11).

24 Nationalisation of private forests led to the vesting of privately owned forests with the Forest Department, through the passing of the Kerala Private Forests (Vesting and Assignment) Act of 1971.

25 In 1800 Francis Buchanan was directed by the then Governor General of India, Marquis Wellesley to travel through and report through the dominions of the then reigning Rajah of Mysore in southern India, and the country acquired by the East India Company in the war with Tipoo Sultan, which included Malabar and Canara.

The erstwhile district of Malabar was a part of the Madras state, which later became a part of Kerala when the state was formed in 1957.

26 'Elephants are of course also still regularly caught in private forests by private janmis, and in many cases, as for instance in the Kollangod forests, form a valuable source of income to the landowner' (Innes 1908:242).

27 When rainfall exceeds the water requirements of a growing crop only for a short period of the year, farmers are left with few options. One is to adjust to the 'natural' season of cultivation by adapting the crops and their dates of sowing to water availability. A second option is by extending the possible season of cultivation by storing it, by obtaining it from sources other than rain or by importing it from outside the area (Athreya, Djurfeldt and Lindberg 1990). All of these options have been experimented with in the region.

28 'and in the Palghat taluk, where the rainfall is less than in the rest of the district, tanks for the storage of water and anicuts over jungle streams are not uncommon' (Innes 1908: 209). The district referred to is the district of Malabar, a district of the erstwhile Madras Presidency.

29 The Palakkad taluk comprised one of the eight taluks of the erstwhile Malabar district.

30 These include the Malampuzha (the first to be implemented in the district during the 1940s), the Kanjirapuzha, Chitturpuzha, Pothundy, Mangalam, Walayar, and the Gayatripuzha irrigation projects.

31 Many other parts of Chittur taluk record a much earlier history of canal irrigation. Canals had been constructed in many parts of the present day Chittur municipality and adjoining areas by the princely kings and large landowning families in the 19th century. These areas therefore exhibited a mix of canal and pond irrigation from much earlier times.

32 Similar to the Palakkad Gap is the Achen Koil Gap or the Aryan Kavu Pass in the Pathanapuram taluk of Kollam district, giving easy access by rail and by road to the adjacent district of Tirunelveli in the state of Tamil Nadu.

33 The Kollengode Block Panchayat consists of five village panchayats (*Grama* Panchayats), which includes the Kollengode village panchayat. The Block Panchayat represents the second-tier in the three-tier panchayati raj system of the country. The Grama Panchayat represents the first-tier and the *Zilla* (District) panchayat the third-tier.

2

Water and Paddy

A Window into Changing Land and Water Management Practices

The objective of this chapter is to illustrate the micro level changes that have come about to land and water management in the panchayats of Kollengode and Elavenchery as a result of state level irrigation and agricultural polices implemented since the 1950s. The purpose of doing so is twofold. By discussing the prevailing state level irrigation and agricultural polices, I attempt to present the logic behind the spread of modern irrigation and the intensification of paddy cultivation in the state. Discussion of these policy objectives partially explains the context in which changes have taken place in agricultural and irrigation practices in the panchayats of Kollengode and Elavenchery, particularly with reference to paddy cultivation.

The second purpose of this chapter is to portray in detail the changes that have come about to land and water management practises in the study area. In order to do so, I have described in detail the traditional pattern of paddy cultivation that was followed in the area prior to the introduction of modern irrigation. The traditional method of land classification, as well as traditional irrigation practices based on small storages in ponds, shallow pits and wells have been detailed upon. This is then followed with a discussion on the changes that were precipitated by the introduction of modern irrigation technology, in particular the dam based canal technology and energised pumps.

Irrigation and Agriculture in India and Kerala

The modernisation of agriculture through the introduction of the Green Revolution package[1] has played a focal role in the agricultural transformation that the country underwent over the past four to five decades. The introduction of the Green Revolution was a post 1965 phenomenon[2], its repercussions being felt in every food-growing pocket of the country. The five-year plans of the country since Independence[3], laid substantial emphasis on agricultural intensification and the use of modern technological inputs in agriculture. The justification of the green revolution package has been mainly on grounds of ensuring food grain production in a food deficit country. In so doing, cereals such as wheat and paddy were given prime importance.

Since the 1950s, creation of modern irrigation facilities, in particular large dam based canal networks has received considerable emphasis across the country. The four major interventions by the government during the 1950s and 1960s that were instrumental in enhancing food grain production were the building of large multi purpose dams to facilitate irrigated agriculture, massive rural electrification programmes, introduction of high yielding varieties of seeds and massive rural credit programmes for capital investments such as well digging (Janakarajan 2002). During the past five decades of agricultural development in the country, the spread of dam based canal irrigation, groundwater irrigation and energised pumping have dramatically increased the quantum of water available for agriculture.

Irrigation in the ecological context of Kerala

Amongst the south Indian states, the state of Kerala can be considered as one of those, which is more or less exclusively governed by a tropical monsoon climate. Rainfall in this region is largely concentrated in two phases, during the southwest monsoon period (June–September) and the northeast monsoon period (October–December). The presence of the Western Ghats as the eastern perimeter of the state exercises the most decisive influence on the topography and climate of the region. The descent of the Ghats into the mainland has led to altitudinal variations manifest in three broad natural divisions in the state, namely the highlands, the midlands and the lowlands, occupying 48%, 42% and 10% of the total geographical area of the state (GOK 1981a). While a mountainous terrain marks the highlands with altitudes ranging between 400–2000 metres, an undulating landscape with alternating hills and valleys, mostly lateritic in nature characterise the midland zone. The lowlands largely comprise of a narrow alluvial coastland. High rainfall, ample sunlight, a variety of soil types and an undulating topography encompassing significant altitudinal variations has led to the emergence of a high degree of bio diversity in the region. These altitudinal and climatic variations enable the cultivation of a range of agricultural and plantation crops ranging from crops like tea, coffee and cardamom in the highlands, tapioca, ginger, pepper and rubber in the midlands and highlands, paddy, coconut and areca nut in the midlands and the lowlands, along with a wide range of fruits and vegetables.

The overall orientation towards irrigation in Kerala has been moulded by the national agenda of providing modern irrigation to enhance agricultural production and productivity. That irrigation facilities were available to only 18.8% of the net cultivated area[4] in the state during the mid 1970s was treated with concern and the State Planning Board strongly advocated for 'greater tempo and speed in the exploitation of the State's irrigation potential' (GOK 1976:19). During the first eight five-year plans (1951 to 1997), the state invested over 14% of the total plan allocation (20750 million rupees) on the expansion of irrigation facilities, spending only a third of the same amount on the development of agriculture and allied sectors (Govt. Rice Commission 1999). The main line of argument was that neighbouring Tamil Nadu, 'with much less water resources', had an irrigated area, which stood, at 41.3% of the net area sown. The seasonal concentration of rainfall in the region was yet another reason cited for the construction and expansion of irrigation facilities in the state. The period of seven months from June onwards receives about 86% of the yearly rainfall, leaving only 14% for the remaining five months. The leanest months in terms of rainfall are January and February, and the monthly variation in average rainfall varies across the state from about 23.5 rainy days in the month of July to 1 rainy day in February (ibid.). It was argued that the shortage of water at the beginning, in between and towards the close of the monsoons created problems for raising two crops of paddy, thereby necessitating the provision of irrigation (Committee on Plan Projects 1960:7). This was considered to be particularly so in the case of the winter (second) crop of paddy cultivated during the months of October–January, whose frequent failure was attributed to a shortfall in rainfall (SPB 1975:2).

The Paddy Focus in Kerala

Enhancing the total production and productivity of paddy has been the central focus of the irrigation and agricultural policies implemented in the state since the 1950s. This was mostly due to the national objective of achieving food self-sufficiency in the post Independence era. In Kerala, achieving food self-sufficiency was equated with enhancing the total output of paddy production[5], justifying investment on irrigation infrastructure. At the end of the first five-year plan in 1956, 69% of the gross irrigated area in the state was accounted for by paddy, a figure that went up to 82% by the end of the fourth five-year plan in 1974(George and Nair 1982:133).

Apart from being the largest beneficiary of the creation of irrigation facilities in the state, the major share of state spending in the agricultural sector was also focused on promoting paddy cultivation. During successive five-year plans, a third of the total plan outlay was spent on the agricultural sector, a major share of which was spent on promoting paddy cultivation, through investment in irrigation, agricultural extension services and a number of special 'package' programmes targeted at paddy such as the Intensive Agricultural Development Programme (IADP)[6], Intensive Paddy Development Programme (IPD), the Yela Development Programme, and more recently the Group Farming Programme (implemented since 1989–90) (Govt. Rice Commission 1999). The Intensive Agricultural Development Programme, a nation wide programme initiated in 1960–61 at the recommendation of the Ford Foundation marked the beginning of agricultural intensification programmes which aimed at enhancing crop yields through the external supply of inputs such as chemical fertilisers, pesticides, high yielding varieties of seeds, and agricultural machinery such as the tractor (SPB 1977:3). A programme to encourage the use of high yielding varieties of seeds was also initiated in the state in 1966 (ibid., 4).

Unsatisfied with the outcome of these package programmes, the state in the fourth five-year plan (1969–1974) outlined a new strategy for intensification by reorganising paddy cultivation on the basis of '*aela*s', in order to enforce uniform cultivation practices on a contiguous stretch of land (ibid., 3). *Aela*s refer to the meandering low valleys found in the midlands of the state, through which flow small streams. The focus of this programme launched in 1971 was to ensure that farmers of each *aela* would act jointly in the procurement and application of inputs as well as in the adoption of improved farm practices (SPB 1977:3). The focus of the subsequently implemented Group Farming Programme in 1989 has also been much the same.

The problematic relationship between irrigation and paddy cultivation in Kerala

Since the lowlands are seasonally inundated and the highlands feature steep slopes, paddy centred irrigation in the state has been mostly focused on the midland zone, which roughly occupies 42% of the total geographic area in the state. Spur hills and low meandering valleys (*aelas*) are the characteristic feature of the midlands, with the valleys levelled into flat fields for paddy cultivation, referred to as '*nilom*' . In situ precipitation, along with surface and sub surface flows from the adjacent lands keep these valley lands moist for most parts of the year, except during the summer months of January–April (Govt. Rice Commission 1999). Gravity flow and terracing of land made agriculture possible in this zone. The valley bottoms are the sites of double-cropped paddy lands, and are referred to as *irippu niloms* (meaning lands where paddy is cultivated twice in a year).

The single cropped or *oruppu niloms* (where paddy is cultivated once a year) are located on the terraced lower slopes, which lie just above the valleys. Unable to retain a significant portion of the rainfall and drainage from upper slopes, these lands can be cultivated with paddy only during the first/autumn crop (April to September), when the southwest monsoon brings abundant rainfall. Single cropped lands or *oruppu nilom*s constituted a quarter to one third of the *niloms* in the state during the 1950s and 1960s (ibid.).

This undulating topography of the midlands therefore imposes a natural constraint on the possible extension of net paddy land[7] in the state. Given the fact that all valley bottoms are already cultivated with paddy, if net paddy land has to increase, paddy has to be cultivated on the sloping lands with relatively poorer water retention capacities. Cultivating paddy on such lands would precipitate a higher water requirement. Moreover, irrigating these lands would require canals to cut across an undulating terrain in order to transport water from distant reservoirs (Santhakumar and Rajagopal 1993). The conventional mode of dam based canal irrigation has therefore been considered to be unsuitable to the specific topographical and climatic features of Kerala (high rainfall, rugged topography and non availability of extensive stretches of valley land) (NCAER 1962 in Santhakumar and Rajagopal 1993). Despite this, the state went ahead with an irrigation strategy that was focused on the construction of large reservoirs and canal networks that aimed at enhancing the area and production of paddy. All the large and medium reservoirs constructed in the Bharathapuzha river basin in Palakkad district, intended to enhance the area under paddy cultivation (S. C. Nair 2004).

In order to increase the total production of paddy in the state irrigation projects aimed at the conversion of single cropped paddy lands into double cropped ones, and double cropped lands into triple cropped ones. The conversion of double cropped land into triple cropped land is argued not to have taken place, as almost all the irrigation projects provide supplementary irrigation only during the winter crop season and not during the summer season when a third crop could be raised (Santhakumar and Rajagopal 1993).

Despite the paddy focus of irrigation policies and the external input oriented approach strengthened by an incentive regime (SPB 1997:10), a decline in area under paddy cultivation began to manifest itself since the mid 70s, a subject of much research and at least a dozen government level enquiries[8] (George and Mukherjee 1986, J. Nair 1981; Kannan and Pushpangadhan 1990:14, S. K. N. Nair 1999:269–70, SPB 1997: 10, Govt. Rice Commission 1999, Kerala Statistical Institute 1994). Figure 2.1 shows the declining trend of area under paddy cultivation in the state.

FIGURE 2.1 Decline in Area under Paddy Cultivation in Kerala

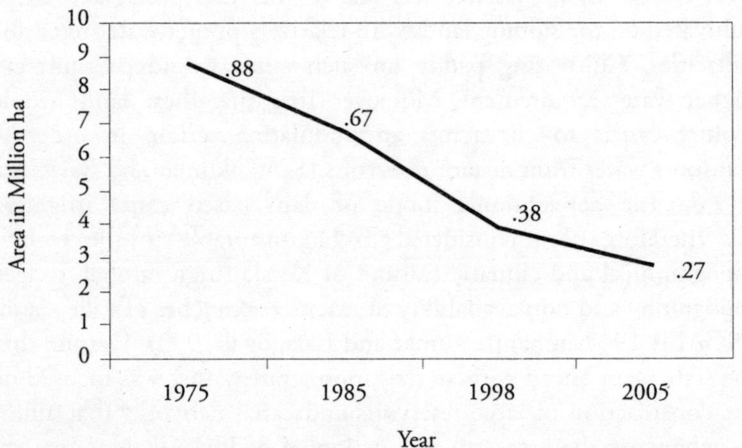

Area under rice production in the state, after having peaked at 0.88 million hectares in 1974–75, with a record annual production of 1.37 million tonnes (SPB 1997:10), began to register a decline. It decreased to less than half in a span of twenty five years (to 0.38 million hectares by 1997–98) (SPB 1997: 10), and further down to a mere 0.27 million hectares by 2005–06 (GOK 2006a). A recent trend analysis by the government reveals that the proportion of paddy land to the total food crop area in the state declined from 48% in 1967–68 to 15% in 1995–96 to 10% in 2004–05 (GOK 2006b). During the last decade alone (1995–96 to 2004–05), a 38.5% decline was registered in paddy grown area in the state (ibid.). Naturally, the state has witnessed a corresponding decrease in the production of paddy, from 1.34 million tonnes in 1981–82 to 0.67 million tonnes in 2004–05, i.e. a more than 50% decrease within a span of 23 years (ibid.).

TABLE 2.1 Area of Important Crops in Kerala for the Years 1961–62 & 2005–06

Crop (Area in Hectares)	*1961–62*	*2005–06*	*% Variation*
Paddy	753009	275742	–63
Tapioca	236776	90539	–62
Coconut*	505035	897833	78
Pepper	99887	237998	138
Rubber	133133	494400	271
Groundnut	15993	3299	–79
Sesamum	11953	600	–95
Pulses	43546	10562	–76
Banana	42693	61400	44

Source: GOK 2006a

* Production of coconuts in million nuts

As the table above indicates, the declining trend in terms of agricultural output was not confined to paddy alone. The crops that have registered an increase in area under production are all cash crops (coconut, banana, pepper and rubber). Of these the crops that registered the maximum increase, viz. rubber and pepper are not irrigated. It has therefore been argued that there is no discernible corresponding effect of irrigation on the total agricultural output in the state as well as on the productivity of agricultural crops on the whole (Vishwanathan 2002, George and Nair 1992, Pillai 1982, Narayana and Nair 1983), the latter being lower than many comparably located regions in the world, and even lower than the Indian average (S. K. N. Nair 1999:271, Kannan and Pushpangadhan 1988, 1990). This has been argued to be true even in the case of Palakkad district where public investment in irrigation has been the highest (Kannan and Pushpangadhan 1989). Vishwanathan argues that even the high cropping intensity in the state cannot be attributed to the irrigation effect, but rather to the prevailing land use pattern that is characterised by a predominance of plantation crops with high density planting (Vishwanathan 2002). It is also argued that there is no evidence to suggest that irrigation has either stabilised the yield of crops, or increased the cropping intensity (Kannan and Pushpangadhan 1989).[9]

Paddy yields do not present an encouraging picture either. According to the Report of the Government Rice Commission, despite the huge investment in major irrigation projects, the productivity of paddy in Kerala has been increasing at a very low average rate of 1.3% per annum[10]. Even this meagre annual increase in productivity was attributed by the Committee to the phenomenon of marginal area going out of paddy cultivation (Government Rice Commission 1999: 12–13), and not due to the provision of irrigation or other enabling factors, a point that has been agreed upon by other researchers as well (Kannan and Pushpangadhan 1990: 9–10). Taluk wise analysis of paddy yields reveal that the yields have been stagnating even in the taluks of Chittoor, Alathur and Palakkad in the Palakkad district, which have recorded the highest yields for paddy (3500–4000 kg per hectare) (ibid.).

It is important to take note of the fact that the decline in area under paddy cultivation as well as stagnating yields is an outcome of a range of factors, and not to the irrigation factor alone. An important factor has been the shift from paddy to the cultivation of commercial crops. During the past four decades, the decline in area under food crops and the simultaneous increase in area under commercial crops such as banana, coconut, areca, and rubber (see Table 2.1) lends support to the argument that the former have been replaced by the latter (Vishwanathan 2002). During the period between 1980–81 and 1997–98, while the share of paddy in the gross irrigated area in the state declined from 73% to 48%, the share of coconut increased from 15.75% to 34.54%, and the share of areca nut and banana has also been increasing (ibid.). This means that irrigated paddy fields have been gradually replaced with coconut, areca, and banana to some extent. An earlier study (Unni 1983) analysing acreage under different crops during the 1960s and 70s came up with a similar conclusion, indicating a trend towards the substitution of paddy with coconut (Unni 1983)[11].

The major reason behind this crop shift is the lack of profitability and high cost of production of food crops including paddy, high wage rates and the labour profitability of competing crops(George and Mukherjee 1986)[12], changed agrarian relations, perceptible decline in the size of operational holdings, non availability and scarcity of water and the relatively arduous and risky nature of paddy cultivation calling for constant care and attention particularly in the realm of water control (Vishwanathan 2002, SPB 1997:10). Yet another important factor behind the decreasing area under paddy cultivation has been the conversion of paddy land for non-agricultural purposes (GOK 2006b) especially for housing plots, thereby reducing the paddy land available for cultivation. The growing density of population and the process of urbanisation in the state have largely been responsible for conversion of paddy fields for building residential and other commercial establishments and related infrastructure including roads (Gopikuttan and Kurup 2004, Vishwanathan 2002). Non-agricultural land use as a proportion of the total geographical area in the state has increased from 5.21% in 1957–58 to 8.19% in 1997–98. (Vishwanathan 2002).

Though the reasons for decline in paddy production in the state are not directly related to the provision or absence of irrigation facilities, the inference that the provision of irrigation facilities has not made paddy cultivation a favourable proposition cannot be ruled out. It has led to the questioning of the continuing investment in large-scale irrigation targeted at paddy despite the declining performance of the crop (Santhakumar et al. 1995). This indicates that the paddy focused agriculture policy of the state needs to be reviewed.

The special case of Palakkad in Kerala irrigation

While the above critique places in doubt the suitability of the conventional large and medium dam based irrigation technology and the continued investment on paddy intensification programmes in the state as a whole, the district of Palakkad however has been a special point of reference in discussions on the irrigation scenario of the state. The specific agro climatic and topographic features of this region have been cited as reasons for the spread of irrigation in the region. As mentioned in Chapter 1, unlike most parts of Kerala, the district of Palakkad, especially the eastern taluks, which are located in the Palakkad Gap, have a history of traditional irrigation. This history of irrigation in the district has however received scant attention in agriculture and irrigation related research in the state. Neither has there been any study of the pond systems and their role in sustaining paddy cultivation in the region. In the post independence era, this district has witnessed the implementation of the highest number of major and medium canal irrigation projects in the state.

The history of modern irrigation in Palakkad dates back to the late 1940s when engineers of the Madras Presidency started planning a few medium and small irrigation projects in Palakkad and nearby areas (Santhakumar et al 1995: A–31). The Malampuzha Irrigation Project located in Palakkad district was amongst the first three major irrigation projects to be implemented in the state. Seven amongst the ten major and medium projects on which work had commenced by 1957 (when the state of Kerala was formed), were located in Palakkad district (ibid.), in the single Bharathapuzha basin[13]. Five amongst these had become fully functional by the fourth five-year plan, i.e. by the 1970s (SPB 1998). It is not clear why there was such a heavy focus on one single district, and one single river basin. Santhakumar and Rajagopalan suggest that like Nanchilnad, the specific agro-climatic conditions of Palakkad (the semi-arid climate here which placed higher water demands) along with the greater expanse of relatively flat plain land led to the 'somewhat successful experience of the storage-based irrigation projects in these two regions' (Santhakumar and Rajagopalan 1993). They argue that the conditions in Palakkad made it more suitable to the technology of large dams and canals, unlike other parts of Kerala characterised by a more rugged topography (ibid.).

Palakkad today is the chief rice-producing district in the state. Of a total of 289974 ha of paddy land in the state, 111029 ha (38.28%) is located in the district, contributing the largest share (39%) towards the total paddy production in the state (GOK 2006b). It also has the highest proportion of double-cropped paddy lands in the state. Of the net area under paddy in the district, 97.1% is double cropped (Kerala Statistical Institute 1994:19). Estimates of reduction in area under paddy cultivation during the past decade (1995–96 to 2004–05) across the state indicate that the decrease has been least in Palakkad district (20% as compared to roughly 70% in the districts of Pathanamthitta, Thiruvananthapuram and Kozhikode) (GOK 2006b).

The above achievements however do not indicate that all is well in the paddy sector in this district. While it is true that Palakkad has registered the least decrease in the area under paddy cultivation, the decrease is not insignificant[14]. During the thirty-year period between 1975–76 and 2004–05, Palakkad registered a 40.04% decrease in the area under paddy.

TABLE 2.2 Area (in hectares) under major crops in Palakkad district during the period between 1975–2005

Year	*Paddy*	*Coconut*	*Banana*	*Other Plantains*	*Pepper*	*Ginger*
1975–76	185182	16994	587	3483	851	383
1985–86	160855	26349	1514	2409	1736	489
1995–96	135630	48336	4413	3409	3460	1239
2001–02	118701	46393	5931	4770	4916	1460
2004–05	111029	55533	10705	6871	7305	969

Source: GOK 2001b GOK 2007a

Crops that registered a substantial increase during the same period were banana, coconut, other plantains, pepper and ginger. The increasing trend in bananas needs to be taken seriously, as banana cultivation is mostly undertaken on paddy land, with banana cultivation resulting in permanent alteration in the land use. In 2005–06, Palakkad ranked second among the districts with regard to area under banana cultivation (GOK 2006a)[15]. While the extent of banana cultivation in the Kollengode and Elavenchery panchayats is limited as per a 2001 survey (State Land Use Board 2001), paddy land is being increasingly used for the cultivation of non-paddy crops. In the Elavenchery panchayat, a total of 103.5 hectares of paddy land (6.2% of the total paddy land) has been converted into plantations of coconut and mixed crops. The latter include mixed cultivation of coconut, areca nut, banana, pepper and vegetables (State Land Use Board 2001). The corresponding figure for Kollengode panchayat is much less, only 0.87%. An additional 10.37 hectares (0.5%) of the paddy land in Kollengode has however been converted into building sites. The higher incidence of paddy land conversion for the cultivation of non-paddy crops in Elavenchery panchayat could be an indication of the inadequacy of existing water supplies in meeting the water requirements of paddy. Elavenchery's location in the tail end of the left bank canal of the Chulliar canal network places it in a vulnerable position as far as timely availability of water is concerned. While many parts of the district have witnessed the conversion of paddy lands into rubber plantations, this has not been the case with the Kollengode and Elavenchery panchayats, owing to the unsuitability of climatic conditions to rubber cultivation.

The data on gross area irrigated (crop-wise) for the district also indicates a declining share of irrigated paddy. The share of irrigated paddy has declined by 7.9% during the period between 1995–96 and 2004–05 (GOK 2001b, GOK 2007a). This could be an indication of the fact that the existing irrigation facilities are not able to meet the water requirements of paddy. The reporting of acute water scarcity during the second crop and third crop of paddy from many parts of the Palakkad district indicates a situation wherein the stabilization of a second crop through assured irrigation is largely dependent on the quantum and distribution of rainfall received during the north east monsoon, with a failure of the north east monsoon often resulting in large scale crop losses even in command areas of irrigation projects (GOK 1999). The replacement of paddy with other crops during the past two to three decades, even in the command areas of major irrigation projects has been viewed as a vindication of the failure of irrigation systems in providing assured irrigation support to originally projected command areas (ibid.). Substantial losses in water by way of conveyance losses and defective distribution systems have also been widely reported (ibid.). In the case of paddy yields, the Expert Committee on Paddy Cultivation attributed the stagnating paddy yields in the rice growing taluks of Chittoor, Alathur and Palakkad in the Palakkad district to the lack of water control (Govt. Rice Commission 1999). Indications of a malfunctioning canal irrigation infrastructure are also reflected in the decline in the net area irrigated by government canals in the district. The net area irrigated (source wise) in the district for the period between 1995–96 and 2005–06, indicates that the area irrigated by government canals have declined by 20% (GOK 2007a; GOK 2001b). In addition, as of 1995–96, only 64% of the area targeted[16] by the Gayatripuzha Irrigation project in which the study area falls was actually irrigated (Vishwanathan 2002).

Reconsidering the Approach to Irrigation

The two pillars of the model of irrigation and agricultural development in the state have been the construction of large-scale irrigation works and irrigation mainly for paddy (Narayana and Nair 1983). The previous sections have illustrated how both of these have not met with their intended objectives. A reconsideration of the irrigation policy is warranted on two accounts. One, the poor performance of the paddy crop during the past four decades demands a reconsideration of a paddy focused cropping pattern. The focus of all irrigation projects on the intensification of paddy has neglected the cultivation of less water intensive crops. This has important implications as far as water conservation is concerned, particularly in regions such as the study area. Santhakumar and Rajagopal have argued that a predetermined paddy dominated cropping pattern not only detracts attention from other food crops, but also necessitates irrigation technology of a particular type and scale, i.e. large storage based canal networks (Santhakumar and Rajagopal 1993). This brings us to the second point, which is the adoption of a single mode of irrigation for the entire state. A reconsideration of the exclusive focus on large-scale canal irrigation is justified not only because of its poor performance, but also with respect to its suitability to the specific topographic and climatic features of the state.

Since the 1980s, the unsatisfactory performance of large scale canal irrigation involving high public investment has led to a greater emphasis being placed on smaller irrigation structures that are based on local agro climatic and geological features (Santhakumar and Rajagopalan 1993, Kannan and Pushpangadhan 1988, Narayana and Nair 1983). It has also raised issues regarding the mode of irrigation planning adopted in the state. Kannan and Pushpangadhan for instance have argued that irrigation planning in the state needs to take into account local specific factors such as crop-mix, soil characteristics, crop specific water demands, topography, soil erosion and siltation problems, rather than rely on large storages alone (Kannan and Pushpangadhan 1988: 39). Interestingly the need to take into account local factors has been voiced with regard to the designing of agricultural intensification programmes as well. The implementation of the IADP, Yela and Group Farming programme for instance has been critiqued for not giving adequate attention to the geographical features of the land and the need for catchment based water management (Govt. Rice Commission 1999)[17]. A critique of the above paddy focused intensification programmes is that they were exclusively focused on paddy lands per se (SPB 1977:47), ignoring the hydrological interconnections between paddy lands and the surrounding lands, particularly so as paddy lands in the state lie interspersed amidst hills and valleys in most parts of the state (Govt. Rice Commission 1999).

Santhakumar and Rajagopal have critiqued the conventional mode of calculating crop water requirements for not taking into account locally available water, especially the run off from surrounding lands which provide substantial amounts of water for paddy in the valley bottoms (Santhakumar and Rajagopal 1993). This is a particular feature of midland Kerala. They further argue that attention needs to be focused on catchment based minor water storages that are more suited to the specific hydrological features of midland Kerala, and on measures that enhance in situ moisture conservation (ibid.). Kannan and Pushpangadhan argue in favour of enhancing public investment in soil conservation, consolidation of holdings, water resources management and timing of irrigation supplies, which has been insignificant, compared to the investment in constructing irrigation infrastructure (Kannan and Pushpangadhan 1988). There is therefore increasing recognition of the need to devise locally relevant water resources management strategies, rather than rely on large reservoirs alone; it is also being recognised that a comparison of the cost effectiveness of various irrigation methods needs to be undertaken before focusing on one particular method (Narayana and Nair 1983). I shall come back to this debate once again after having discussed the local forms of agricultural and irrigation management in the study area.

The Changing Patterns of Agriculture and Irrigation in the Varayiri Watershed

This section analyses the micro level impact of the above discussed irrigation and agricultural policies in the study area. I begin with a description of the traditional agriculture and irrigation practises in the area, followed by the changes brought about with the introduction of modern canal networks and energised pumping devices. While describing the traditional system of agriculture and irrigation in the area, I focus on the ways in which they ensured an optimum use of available water supplies.

Land use pattern

The undulating terrain of the Varayiri watershed is reflected in alternating hills and valleys that drain water into small streams that make up the Varayiri stream. This is a typical feature of the midland zone in the district and the state.

Agricultural land comprises a major part of the total geographical area in the watershed. Land use data pertaining to the panchayats of Kollengode and Elavenchery in which the watershed is located is an indication of the land use pattern within the watershed. Agricultural land comprises 66.72% and 61.46% of the total geographical area in Kollengode and Elavenchery panchayats respectively. Of the total agricultural land, paddy land occupies 65.8% and 79.30% in both the panchayats respectively (Kerala State Land Use Board 2001). When expressed as a percentage of the total land area, paddy land in itself occupies 43.9% and 48.74% of the total land area in Kollengode and Elavenchery panchayats respectively.

TABLE 2.3 Land Use Classification in the Panchayats of Kollengode and Elavenchery

Land Use	*Kollengode*		*Elavenchery*	
	Area in ha	*%*	*Area in ha*	*%*
Agricultural land	2767.97	66.72	1977.75	61.46
Forest land *	712.01	17.16	902	28.03
Land under mixed trees **	350.84	8.46	100.75	3.13
Wasteland	26.47	0.64	77	2.39
Land under buildings	60.99	1.47	103.75	3.22
Rocky area	67	1.61	4.75	0.15
Water Bodies	163.41	3.94	52	1.62
Total	4148.69***	100	3218	100

Source: Kerala State Land Use Board 2001

* It needs to be noted that the area shown as forestland does not comprise of good forest. It only indicates the extent of land with the Forest Department and does not indicate the existence of undegraded forest.

** Land under mixed trees includes small private plantations of teak, mango and other trees.

*** While the actual land area of the Kollengode panchayat is 4933 ha, an area of 784.31 ha was excluded from the survey conducted by the State Land Use Board. As a result, the above-mentioned percentages have been calculated for the surveyed area, i.e. 4148.69 ha.

Prior to the introduction of modern irrigation and the intensification of paddy cultivation in the area, the valleys were reserved for the double cropping of paddy, the lower slopes for dry crops and a single crop of paddy, and the higher slopes covered with trees, some dry crop cultivation, and were also sites for human settlement. Hence, a wide mix of trees in the higher elevations, and crop land in the lower slopes formed the land use picture of the watershed. The prevailing system of agricultural land classification was based on elevation differences and soil types. Cultivable land in general is comprised of *parambu* and *padam.* Parambu land is located on the higher slopes, partly occupied by houses and a wide mix of trees. P*adam* refers to paddy land in the lower slopes and valley bottoms. Padam land, was further divided into two broad categories, *potta* and *kalayi.* The former refers to single cropped paddy land located on the lower slopes, while the latter refers to double or even triple cropped paddy land located in the valley[18] (see Figure 2.2 for an illustration of the same). This classification was based on the availability of moisture for growing paddy under natural rainfall conditions.

There are finer variations within each of these categories, which is captured in the local terminology. These fine categories indicate the degree to which fields retain moisture. While kalayi lands refer to low lying lands, located in the valleys, *elamkalayi* (*elam* meaning light) refer to lands, which are low lying, but not located right at valley bottom. *Adikandam* on the other hand (*adi* meaning lowest, and *kandam* meaning paddy field) refers to kalayi lands, which are located at the lowest part, or at the valley bottom. *Tazhathe kandam* is yet another term that is used to refer to low lying paddy fields. Both the above types retain maximum amounts of water and soil moisture. In contrast, water is drained off sloping potta lands much faster. Amongst potta lands, *metu kandams* refer to land lying on the higher slopes, which retain very little water. One can hear farmers narrate the number of times they have had to irrigate these fields due to poor water retention. In general, the surface of potta lands dries up much faster leading to lower soil moisture levels.

In the traditional pattern, only one crop of paddy was raised on potta lands during the first crop season (April–September). These lands were more often than not sown with a variety of less water demanding dry crops during the months of October–December. Farmers today recall how they grew black gram, sesame (*ellu*), groundnut, and horse gram on potta lands during the second crop season, prior to the introduction of the canal network in the late 1960s. Velayudhan, who comes from a hitherto large tenant background, recalls how his father used to cultivate black gram on two hectares of single cropped land during the second crop season. Black gram was preferred as it was a good fertiliser for the fields. This tradition of dry crop cultivation is borne out by Innes's remark in 1904–05,

> In the eastern *desams*[19] of the Palghat taluk, that adjoin Coimbatore district every variety of grain grown in the latter district is found, groundnut, *varagu*, *ragi*, blackgram, *chama*, *cholam*[20], horse gram and even cotton. The cultivation of the first is extremely profitable and blackgram is a valuable crop in South Palghat (Innes 1908:216).

Farmers recall that the mist during the months of October–December was conducive to the growth of various varieties of beans and gram. The periodic rotation of crops with nitrogen fixing ones such as various varieties of beans is considered to improve soil fertility as well (Hanks 1972). This has been a feature of most regions that face seasonal shortages of water.[21]

It is important to note that the land cultivated by farmers comprised of a mix of land types including parambu, potta and kalayi fields. The three together, along with the forests in the surrounding areas met the resource requirements of the farmers. In the context of cattle raising, Buchanan for instance notes 'Every man who occupies rice-land has a certain part of the high land attached to it for pasture; and to this he has an exclusive right, without paying rent. . .' (Buchanan 1870). Today, this mix of land types within one single holding can be observed only in the case of the larger holdings. 4.8 hectares jointly owned by Krishnankutty and his brothers in Kannankulambu for instance consists of 3.2 hectares of paddy land, and 1.6 hectares of parambu land, which has now been converted into mango plantations. Similarly Raman Nair owns 3.2 hectares of paddy land, and about one hectare of parambu land. They however are exceptions, the majority of holdings are much smaller in size, consisting of paddy land alone, and in many cases of either kalayi or potta land alone.

Ponds and the flow of water

Like many other parts of the Chittur taluk, ponds formed an important element in the landscape of the Varayiri watershed, with more than a hundred and fifty ponds spread out over the watershed, which covers an area of 1450 hectares. The southern (the highest) ridge of the watershed runs at an average elevation of 150 metres from east to west, the highest point being the Cheerani hill at 163 metres. The Varayiri stream meets the Gayatri River at an elevation of 128 metres. Thus this small stream (only 15 kilometres in length), traverses through a terrain that exhibits no significant variations in gradient. It is therefore surprising that 150 ponds have been carved out of such a terrain, and that too within a relatively small area, there being a pond in every 100–200 metres. Such a high concentration of ponds is not unique to this area alone, but a common feature in the adjoining panchayats of Mudalamada, Peruvamba and Pudunagaram.

There is a lack of clarity regarding the age of these ponds. Most farmers from the area report that the ponds have been here since the time of their grandfathers and even before. Fifty-year-old Krishnadas from Manalipadam recalls his grandfather telling him that ninety people worked on the construction of the Perinkulam in Manalipadam. Older farmers recall that it was the large landowning class in the pre land reform era that took the initiative of constructing ponds, in order to provide irrigation to the paddy lands.

In the absence of any significant fall in gradient, the creation of such a large number of ponds suggests a thorough understanding of the minor undulations and the flow of water in each micro catchment. Interrupting and reorganizing the overland portion of the hydrologic cycle in order to secure a supply of irrigation water is the underlying principle of these traditional water management structures (Wilken 1974:50)

Careful diverting and directing of rainwater is essential to the filling up of ponds. The technology employed has been simple, consisting of an embankment built across the line of drainage so as to hold back surface run-off. The embankment is known as '*varambu*', larger ponds having a higher and wider varambu. The main bund

PHOTO 2.1 The Pond Bund (*Varambu*)

is then supported by two other bunds on either side. On the remaining side there is no bund, as it lies along the slope. Sluices were located on the main bund in order regulate the outward flow of water.

The technology involved in the construction of these ponds is similar to that employed in the construction of ponds elsewhere in southern India, or the *ahars* (similar water harvesting structures) in South Bihar. The following definition of the *ahar* makes it seem as though it was written for the pond

> . . .south Bihar has a marked slope from south to north . . . using this terrain condition, an *ahar* is made by erecting an embankment of a meter or two in height on the lower ground . . . from the two extremes of this embankment two other embankments are constructed so as to project towards the higher ground . . , gradually diminishing in height as the ground level rises, and ultimately ending at the ground level . . . the fourth side is left open for drainage water to enter the catchment basin following the gradient of the country (Sengupta 1996:175).

The same principle guides the construction of ponds of south India, referred to as the *ery*, as well, where 'the bund surrounds the water on three sides. The fourth side is open to the catchment from which water flows down to collect in the Ery' (Mukundan 1996:71). The water thus collected is then distributed through sluices fitted on the embankment. Systematic storage of runoff water, controlled release and distribution of gravity are therefore the three important features of this age-old irrigation technology (Von Oppen, Subba Rao and Engelhardt 1986).

The average area of the ponds in the watershed is between one and three acres, though there are some larger ones covering an area between four and eight acres. The average depth of most ponds at the centre ranges between 6–10 feet. Incidentally, two of the bigger ponds in the watershed are both named Perinkulam, meaning 'big pond'. In the case of Perinkulam in the Manalipadam area, the pond covers an area of six acres, with a depth of 6–8 feet at the centre.

If a person were to stand at the centre, he would be well under water. The depth at the sides of the pond is lesser owing to the sides being silted up. However, this is a relatively better maintained

Photo 2.2 A Pond with palymra trees on its bund

pond. Nager potta on the other hand represents many of the smaller ponds in the watershed. It comprises an area of just sixty cents (0.24ha), and its bed has been substantially silted up. As a result, it resembles a very shallow bowl with paddy fields all around.

The drainage into each micro catchment in the watershed is marked by a series of ponds along the drainage lines. This is evident from the fact that the origins of each of the four small streams that feed into the Varayiri lie in or below a perennial pond. Each pond is given a name. As mentioned in Chapter 1, ponds are commonly referred to as kulam or eri in the area. While the names of bigger ponds are familiar to most people, the names of the smaller ponds are known only to the farmers in the immediate vicinity. Not just ponds, paddy fields too are given names. Krishnankutty, from Kannankolambu, who owns lands in the *ayacut* (command area) of the Pedamuri kulam, listed out names of some of the fields in its ayacut. The field immediately below the pond is referred to as *erikandam* (*eri* meaning pond and *kandam* meaning field); the one with a large mango tree by its side is referred to as *moochikandam* (*moochi* meaning mango tree and *kandam* meaning field); another is *challittakandam*, referring to the field in which the first channels would be ploughed marking the commencement of the first crop in the month of April; *vadake kandam* referring to the field that faced the north (*vadake* means north); *kaatukandam* referred to the field where the soil resembled forest soils (*kaatu* meaning forest); *adikandam* referring to the most low lying field in the ayacut, and there was one called *kunukan* because it was shapeless!

Box 2.1 Ponds and Streams

The combined flow from the Kannankulam, Kannankolambu kulam and the Pedamuri kulam on one arm, along with Velankulam and Pathykulam can be considered as the origin of the Varayiri stream, collecting drainage from the southwestern parts of the watershed. The drainage from the Cheerani hill (its northern and eastern slopes) emerges as another sub stream after having flowed through a series of ponds. The Maripadam *todu* (todu meaning stream) is yet another stream that takes off from a pond, the Maripadam pond. While the above mentioned streams collect drainage from the southern parts, that from the north eastern part flows through two other streams, one originating from Odungatuchera (chera refers to a pond that is not very deep) and the other from near Poricholam *kulam*. The drainage from the northern parts of the watershed, including that from the Kollengode town joins the Varayiri through a smaller stream called the Kallantodu. Having travelled for about three kilometres after being joined by the Kallantodu, the Varayiri is joined by the Pulikalpotta stream, which collects drainage from the mid reaches of the watershed. Soon after this junction, the stream cuts the Trichur-Pollachi main road at Valluvakundu, and joins the Gayatri River.

Source: Field Notes, November 2001

Ponds and land use

The catchment of the ponds in the upper and middle reaches of the watershed mostly comprise of parambu lands. This is the case with the two big ponds, the Gramakulam and the Velanganpadam kulam, depicted in Figure 2.2. The area shown under mixed trees in this figure (which also includes coconut plantations), is located on the elevated portions of land, which primarily consists of parambu land. Until about thirty years ago, these lands had fairly good tree cover. Some parts of the parambu land were wild, comprising of *valartukadu* (small patches of dense vegetation), and other parts consisting of trees largely grown for human use. The valartukadu normally enclosed a small *kavu* (a place of worship), where deities of snake-gods were kept. These are referred to as '*paambum kavu*'. Even today, there is one such paambum kavu in the catchment area of the Velanganpadam kulam. Similarly, the catchment of the Perinkulam in Manalipadam, included parambu land on which grew a wild patch of trees, now a part of a cemetery. Farmers from nearby recall cultivating groundnut on this parambu land. Kateri (*Katu* means forest and *eri* means pond) in Pulikalpotta for instance was so called because it was filled by *katuvellam* (meaning water that flows down from the forest).

Housing clusters, usually inhabited by distinct caste groups, were also scattered on hillocks throughout the watershed, and water flowing through these settlements (kudiyirippu or taras as they were referred to) was a major source of inflow into the ponds. As can be seen in Figure 2.2, the housing clusters are located in the upper reaches of the micro catchment, and water flowing through drains into the two ponds. The catchment of ponds in the lowest reaches comprises mainly of paddy land (see the ponds located amidst kalayi fields in Figure 2.2).

PHOTO 2.3 A pond in the valley bottom, amidst paddy fields

FIGURE 2.2 The Catchment and Command of the Velanganpadam Pond

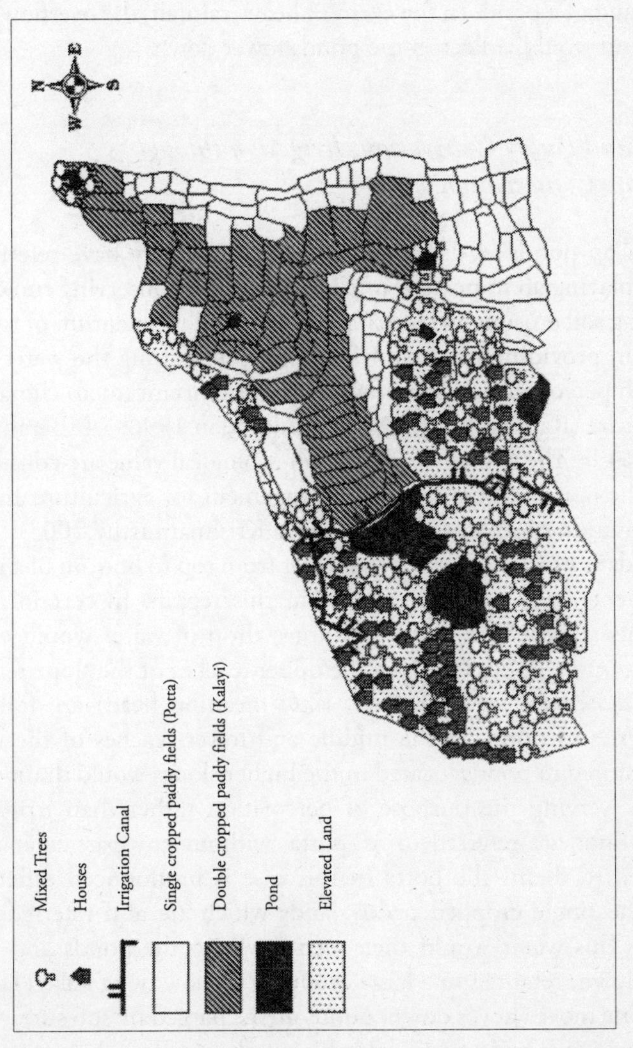

Source: Adapted from Kerala State Land Use Board 2001

As one moves down the slope, the ayacut of one pond would form the catchment of the one lower down. Hence, one can say that almost all the land in the watershed was located either in the catchment or command of a pond. In the event of heavy rainfall, the overflow from one pond would collect in the ponds lower down.

Ponds and Paddy Cultivation: Irrigation through Moisture Conservation

Studies on ponds in different parts of the country have referred to ponds playing an important role in storing and conserving run off, in reducing soil erosion by reducing the pace and momentum of run-off water, in providing low cost irrigation, in enriching the water table through percolation and in providing an important micro climate for agriculture (Raju and Shah 2000, Mukundan 1996: 74, Barah et al. 1996: 149). The two key functions of ecological value are considered to be the optimisation of water management for agriculture and the minimisation of soil loss (Gunneli and Krishnamurthy 2003).

Ponds storing and conserving water from top to bottom of a micro catchment are a common feature in this region. In certain micro catchments, it would seem that not a drop of water would escape flowing into a pond. Ponds in the upper reaches of the slope referred to as '*talakulam*' or '*talachera*', (*tala* meaning head) are followed by a series of ponds at the middle and lower reaches of the slope. Water stored in ponds located in the higher slopes would drain down quickly, serving the purpose of percolation rather than irrigation. Such ponds are referred to as 'potta' without any particular name attached to them (the potta in this case is pronounced differently from the single cropped paddy lands which are also referred to as potta). This water would then seep down to the ponds and fields lower down, retained to a large extent in the low lying kalayi lands.

As one moves lower down, ponds are recharged by sub surface run off from ponds and paddy fields higher above. Though many ponds lower down dry up in summer, perennial springs are found in some of the pond beds. In summer when the pond is more or less dry, on digging a few feet at certain points in the pond bed, sub surface flows are found. These are important sources of water for people in the vicinity. Ponds in the valley bottoms, are surrounded by paddy fields all around, and are often bordered by small streams too. These ponds are found to have slushy, clayey beds. Efforts to deepen such ponds have failed many a time, a thick slush, called '*cheru*' keeps oozing out. Farmers describe it as thick curd, which refuses to be confined.

The system of agricultural land classification discussed above (with the classification of kalayi and potta lands) was defined by the drainage pattern of each micro catchment. The main bund of the pond was constructed across the line of drainage, with kalayi lands located immediately below the bund and along the drainage lines (see Figure 2.2, where the kalayi fields lie along the drainage lines). Water stored in the pond therefore helps to maintain soil moisture levels in the kalayi fields below for a longer period of time. Kalayi fields also get the '*ootu*' (sub surface seepage) from water stocked in the pond. The increased wetness in these fields, imparts additional coolness (*tanapu*), which helps the paddy crop to hold on till the next spell of rain or the next turn of water release from the pond or the next turn of water release from the canal network as is the present case. As far as ponds are concerned therefore, storage as well as recharging of sub surface flows of water are equally important functions.

Given the critical position occupied by ponds, filling them with rainwater was an activity that was highly prioritised. Owing to the high frequency of ponds, most of them have to be filled up by drainage from a relatively small catchment area. Most of the older farmers who have been farming here for at least the past three to four decades recall the urgency with which efforts were made to fill up the ponds, so much so that there would be conflicts over water being led to certain ponds and not to others. The drainage paths by which rainwater was led into each pond were clearly identified and cleaned before the monsoons.

Sub surface recharge is not only manifest in additional wetness in low-lying fields; it helps in recharging other small water sources in the catchment as well. Apart from ponds, the shallow pits referred to as kuzhis (meaning a depression) have been dug in the valley bottoms, tapping into perennial sub surface flows. These kuzhis are well placed, usually at the end of the slope, into which filters sub surface flows from the surrounding slopes. As many farmers narrate, "we pump it dry tonight and by tomorrow, you can see water trickling in". The water level in kuzhis is also reported to rise when water is being supplied through the canal system in the upper reaches of the watershed. In the past, they were an important source of drinking water for cattle. Though small, pumping from kuzhis provide relief when other sources fail to deliver. This is typically observed towards the end of the second crop period, when the rains have withdrawn, and when supply through the canal network has stopped. Kuzhis are also used to irrigate coconuts or a small patch of vegetables.

In the past when pumping technologies were not in use, manual lifting of water was the only method. Hence, kuzhis were not viewed as a significant source of irrigation. With the availability of energised pumping facilities, and with fragmentation of holdings, kuzhis are becoming more critical, despite providing only limited amounts of water. Many such small kuzhis have therefore been deepened and widened resembling square shaped wells. In some of the large holdings, the land is irrigated by a network of ponds, kuzhis and wells. Kuzhis and wells that are located in paddy fields or close to a pond have a higher chance of being perennial as the water in the pond recharges the water level in the former. Many panchayat wells dug for drinking water purposes are located by the side of ponds. The recharging of wells and kuzhis by water stored in ponds illustrates the water optimisation effect of ponds (Gunneli and Krishnamurthy 2003).

Apart from kuzhis and wells, another system to tap water was the *chera*. A chera refers to a paddy field that is cultivated with paddy only during the first crop. During the second crop the field is stocked with water. The bund of the chera, which lies in the direction of water flow, is higher than the bunds around an ordinary paddy field. The supporting bunds on either side are also slightly higher, though not as high as the main bund. The high bunds of the Tumbikod chera is reported to have led to substantial amounts of water collecting behind it, requiring water to be baled out during the first crop season when it was cultivated with paddy. Most cheras are seen to be located in the lower reaches of the watershed, many a time by the stream. Being low-lying lands, these cheras were able to collect substantial amounts of water in the second crop season, which were then used to irrigate nearby fields. This system is no longer in use as the introduction of the canal network and the possibility of additional supplies of water during the second crop season led to cheras being used as double-cropped paddy land. A total of ten cheras in the watershed have been converted into paddy fields since the introduction of the canals. This enumeration has been based on reports from farmers, as identification of hitherto cheras is not as easy as the identification of ponds. Hence the number could as well be more.

Hence, a network of ponds at different levels, along with streams, wells, kuzhis and cheras tapping into surface and sub surface flows, scattered amidst the agricultural land of the Varayiri watershed created a moisture regime that was conducive to paddy cultivation. Storing scarce water in ponds and cheras, and recharging water sources in the immediate ayacut such as wells and pits seems to have been the underlying design. A mutually interrelated system is seen to exist, with the ponds playing a pivotal role.

Water distribution from the ponds

Mendis insightfully describes how the invention of the sluice marked a turning point in ancient irrigation, enabling the construction of storage reservoirs. He differentiates between water management in space as represented by river diversions from water management in time as represented by storage reservoirs. Without this device for control and issue of water, the construction of storage reservoirs would have limited value (Mendis 1999:58, 66), and could even be 'dangerously unfeasible'. As the sluice stands for water control, it is considered to symbolise the main predicament of south Indian agriculture, which is to control water to extend the season of cultivation (Gunneli and Krishnamurthy 2003). Interestingly there is close similarity in the Sinhalese word for sluice (*sorowwa*) (Mendis 1999:58) and the Malayalam word for the same (*ovu*).

Sluices placed on the main and side bunds ensured supply of water from the pond to the surrounding fields. Water was supplied through sluices placed at the higher and lower levels, *mele ovu* and *kizhe ovu* as they are called (*mele* meaning higher and *kizhe* meaning lower). The high level sluice was usually placed on either of the two supporting bunds of the pond. When the pond was full, water would flow out through the mele ovu to the fields located on the two sides of the pond. These fields (mostly potta fields), were located at par with the water level in the pond or slightly above. The water let out to these fields would gradually seep into the kalayi fields below. Once the water level receded, water would not flow out of the mele ovu. In the period before the introduction of pumping devices, once water receded below the level of the mele ovu, it had to be lifted out manually to the higher fields provided they were in possession of a water right. Buchanan notes of this practice:

> In some places, where there is not a sufficient level, the superfluous water is thrown off by a basket suspended between four ropes, and wrought by two men; a manner of raising water practised in China, as well as in every part of India (Buchanan 1870: 69).

Manual lifting of water however was resorted to only during critical periods of the first crop. The lower sluice, what was known as the kizhe or *tazhe* ovu was located on the main bund, and opened out into the fields lying directly below the pond, i.e. along the natural drainage. Hence the kalayi fields lying below the pond had access to a larger volume of water. The sluice opening into the kalayi fields below was technologically more sophisticated. Cylindrical pipes, about 3–4 feet long were placed at the bottom of the main bund. Initially these pipes were made of the bark of the palymra tree. The insides of the palymra tree were scooped out, and the pipe like structure obtained was referred to as '*panayude pati*', '*pana*' meaning palymra and '*pati*' meaning drain through which water can flow. The outlets of these pipes were plugged, functioning like a valve, when the supply of water had to be terminated. In order to close and open the sluice, it was necessary for somebody to get into the standing water in the pond. Later, these sluices began to be made in clay by the traditional potters of the area.

The system of water distribution from ponds adheres to the drainage features in each micro catchment, thereby adhering to the kalayi-potta classification. The kalayi fields get the maximum benefit of the water stocked in the pond, both in terms of seepage as well as quantum of water supplied through the sluices. Since the potta lands retained a lesser amount of soil moisture, they were cultivated only during the first crop season. By limiting double cropping of paddy to kalayi lands that were insured by the water stored in the ponds, the risk of crop loss was reduced. This also ensured efficient use of water, for irrigating potta lands would be a gamble. In the event of a poor monsoon, when the water in the ponds was not enough for even the kalayi fields during the second crop, not all the land in the pond ayacut would be cultivated, and a discretion was exercised by the farmer whereby only those kalayi lands which retained the maximum amount of water were sown with paddy, the others would be abandoned. Such a decision however was easier to arrive at as only one or two tenants, as compared to the current situation where a number of farmers own different parts of a pond ayacut, would cultivate the entire ayacut of a pond.

Fertilising the water and the fields

The role of ponds in the fertilisation of fields is also significant. Once in every few years, accumulated silt on the pond bed was taken out in baskets and thrown on the fields below. This silt, referred to as *cheru*, was valued for its fertility. By interrupting overland flow and arresting the sediment behind the pond bund, ponds helped to retain the soil that is lost through agriculture in the upper catchments, with the fine textured pond bed sediment being restored to the fields (Gunneli and Krishnamurthy 2003). This phenomenon is also considered to partially restore the balance of the soil texture and optimise on site fertility (ibid.).

Cheru was taken out when the ponds were dry in summer. The work of desilting was normally entrusted to members of the P*arayar* community. The P*arayar* in the area reside in distinct taras (housing clusters) at Payilur, Elavenchery, Vattekad and Kollengode. They were paid according to the number of baskets of silt that was lifted out. They were paid both in money and in kind. The Parayar were also called for strengthening the bunds of the paddy fields and ponds during the summer season, prior to the commencement of the first crop.

As mentioned earlier, except for ponds located in the middle and lower valleys, the catchments of most ponds comprised of parambu land. This also included land for grazing cattle. Hence, the first rains would wash in organic material consisting of decomposing leaf litter and cow dung into the ponds, which would then flow down into paddy fields. The role of the pond in fertilising the water through the accumulated silt on its bed, as well as flushing in the organic material from its catchment was important in maintaining fertility of the paddy fields. Farmers recall how they would try to get as much as possible of this '*pudu vellam*' (first showers) into their pond, and not let it flow down into another. Interconnections between forests, land use and paddy cultivation have been observed in the case of ponds of Tamil Nadu and Karnataka (Mukundan 1996: 74), as well as in the Dry Zone of Sri Lanka (Ulluwishewa 1991: 104). In the latter case, Ulluwishewa describes in detail how the village forest in the catchment area and the perennial tree cover on the cultivated area helped in maintaining the productivity of the cultivated area in the village.[22]

In Geertz's detailed and nuanced analysis of wet rice cultivation in Indonesia, he explains why the rice crop is more dependent on the 'medium' (i.e. water), rather than on the solid surface in which it is rooted (the substratum) (Geertz 1963: 18–19). In comparison with shifting cultivation, where the decomposition of organic material supplies the necessary nutrients to the rice crop, fixed and settled fields are considered to rely heavily on water, the 'life-bearing brew' (Hanks 1972). The rice on settled fields is considered to depend on the water to bring in nutrients, with organic matter and minerals being brought in when streams pass through forests and villages on their way to rice fields (ibid.). This explains why the '*pudu vellam*' was such a valued resource.

The seasonal agricultural calendar

Paddy is a seasonal plant in its maturity patterns. As a result the planting calendar and timing of agricultural operations become critical for many varieties to enable them to take advantage of micro climatic factors that cause the grain to mature (Spencer 1966 in Schneider 1995). Similarly, the length and timing of the dry and rainy seasons as well as the timing of the onset of the rainy season exercise a significant influence on plant growth (ibid.). The seasonal nature of plant growth therefore indicates the long process of fine-tuning that must have taken place in order that rice cultivars become adapted to their micro climates. The large number of local varieties found in each rice-growing pocket is indicative of the above process. Pandian in his study of irrigation and agriculture in Nanchilnadu has described the synchronisation of the crop cycle to the monsoon patterns (Pandian 1990). Farmers were observed to choose different varieties of paddy for each of the monsoon periods, with the time structure of the selected paddy varieties corresponding with that of the monsoons.

Such a synchronisation is reported to have been widely practiced in the study area prior to the introduction of modern high yielding varieties, and before the provision of modern irrigation removed seasonal considerations. Apart from a restriction on double-cropped paddy land, the tuning of the cropping calendar to the monsoons enabled farmers to minimise the demand for supplementary irrigation. The autumn crop of paddy (the first of the two crops) commenced in April–May just before the pre monsoon showers. It is referred to as the *Kanni* crop, as it is harvested in the Malayalam month of Kanni (coinciding with the months of August–September). Since the crop cycle commenced towards the end of the dry season, transplanting was not resorted to for want of adequate amounts of water. This was the time of the year when ponds and streams ran dry in the region. Broadcasting, defined as the scattering of seeds on the surface of a field (Hanks 1972) took place while the fields were still dry, the scattered seeds being lightly covered with dust (ibid.). This form of planting was referred to as '*podi veda,*' meaning sowing in dry soil (*podi* meaning dust and *veda* meaning the act of sowing).

Broadcasting was mostly conducted towards the middle of April, coinciding with the New Year festival of *Vishu*[23], when a few showers are anticipated. During the few weeks preceding Vishu, one could see ploughed fields spread out all over the area, as though in anticipation of the rains. The young crop of paddy holds on till the onset of the southwest monsoon in early June with the aid of intermittent pre monsoon showers, which are common in the months of April and May.

The second crop of paddy was more difficult. For one, the northeast monsoon was not as abundant as the southwest. The seasonal break-up of annual rainfall in Table 2.4 indicates the significantly lesser amount of rainfall received during the summer months of January–April (approximately 5% of the total annual rainfall). Since the northeast monsoon would usually withdraw by December, farmers were very anxious to harvest the second crop of paddy by then.

TABLE 2.4 Break-up of annual rainfall received (in mm) at Kollengode during the southwest monsoon period, the northeast monsoon period, the summer months and the pre monsoon period between 1999 and 2002

Year	*South West Monsoon (June–September)*		*North East Monsoon (October–December)*		*Summer Months (January–April)*		*Pre Monsoon (May)*		*Total Annual Rainfall*
	Amount	*% of annual rainfall*	*Amount*	*% of annual rainfall*	*Amount*	*% of annual rainfall*	*Amount*	*% of annual rainfall*	
1999	942	63.20	359.8	24.14	57.6	3.86	131.2	8.80	1490.6
2000	1222.4	84.69	123.6	8.56	91.6	6.35	5.8	0.40	1443.4
2001	1093.6	67.65	358.8	22.19	81.4	5.03	83	5.13	1616.8
2002	934.3	70.92	243.3	18.47	46.0	3.49	93.8	7.12	1317.4

Source: Records maintained at the rain gauge station at Kollengode Railway Station

Transplanting, the technique of moving young rice shoots from a nursery and setting them out in a larger field to grow and produce grain (Hanks 1972) was central to the second crop. Raising a nursery in advance helped farmers to commence the cycle early, resulting in an early harvest. Farmers today vividly recall how paddy seedlings for the second crop were raised in seedbeds much before the harvest of the first crop. They were raised in the months of July–August, the months of heaviest rainfall. Unlike the present, these seedbeds were not raised on paddy fields, but on parambu land, and were referred to as *naatupottas* or *naatukadu* (*naatu* meaning saplings and potta meaning high land). These naatupottas were in most cases a part of the parambu land.[24] Raising the nursery prior to the harvest of the first crop ensured that the rice shoots would be ready for transplanting as soon as the fields were harvested. This simultaneous process of harvesting and transplanting marked a very hectic period for farmers and labourers, particularly so for the latter. Ponds played a critical role at the time of transplanting, especially if water was not present in adequate amounts in the fields to be planted.

When the traditional seed varieties were in use, it is reported that seedlings would be retained in the nurseries for as long as 50 days, viz a viz the current practice of transplanting within 22–30 days. This was to ensure that the seedlings would have to spend a much lesser time in the open fields, thereby reducing their water requirement. '*Naatinnu moopu koota*' was the norm, meaning 'to increase the maturity of the saplings'. To the contrary, the current practice of transplanting within 30 days implies that the saplings have to spend roughly 90 days in the open field. In the traditional method, as soon as the harvest of the first crop was over, the saplings would be plucked and tied up in bundles, and thrown into wet, ploughed fields. Since an early harvest of the first crop was essential for an early beginning of the second crop, the maximum maturity of the varieties used for the first crop did not exceed 100–110 days. Some of the traditional varieties used during the first crop would even mature within 90–100 days.[25] The time structure of the selected paddy varieties therefore corresponded with the monsoon cycle (Pandian 1990). *Vrichikapandi* for instance, a seed used during the second crop was so named as it was harvested in the Malayalam month of *Vrichikam*, which coincides with December. Setting the cropping calendar to the rainfall pattern therefore maximised the chances of getting two crops of paddy on kalayi lands.

Changes in the agricultural milieu[26]

This section discusses the changes that have taken place in the agricultural milieu of the Varayiri watershed. Some of the key changes include the introduction of the canal network, the energised pumping devices as well as the introduction of high yielding varieties of seeds and chemical fertilisers.

The Superimposition of the Canals on the Waterscape of the Varayiri

The Gayatri Irrigation Project comprising of the Meenkara and Chulliar reservoirs and the associated canal networks became functional during the late 1960s. The Left Bank Canal (LBC) of the Gayatripuzha systems takes off from the Meenkara reservoir, and later links up with the Chulliar reservoir and continues as the LBC of the combined system (see Fig. 5.1 in Chapter 5). A total of ten distributary canals take off from the left bank canal, with the Chulliar reservoir supplying water to eight of them. The Varayiri watershed falls in the command area of the last four distributary canals, viz. the Kollengode, Payilur, Peringotukavu and Karinkulam distributary canals. It therefore falls in the middle and tail reaches of the left bank canal. Field channels taking off from the main and distributary canals supply water to paddy fields. These field channels are referred to as CADA channels as they were cemented by the Command Area Development Authority (CADA) under the Command Area Development (CAD) programme initiated in 1985.

This network of main and distributary canals along with the field channels was imposed upon the agricultural landscape marked by parambu, potta and kalayi lands, along with ponds, cheras, kuzhis and streams. The left bank canal traverses along the southern high ridge of the watershed, coinciding with the ridgeline in many parts. (see Map 1.3 in Chapter 1). The distributary canals and field channels also traverse through the relatively higher portions of the watershed. As a result, water supplied through the canal network is easily diverted into ponds located further below. Some amount of the water supplied to paddy fields also seeps into the streams below. Since 1970, the water sources within the watershed (ponds, wells and streams) have been recharged by the water supplied through the canal network during the months of October–January.

Introduction of Energised Pumping Devices

The introduction of energised pumping radically transformed water use patterns from surface and groundwater sources. The kerosene/diesel pumps which made their appearance around the 1960s were soon followed by the electric pump sets in the 70s. These water-lifting devices are used to lift out surface water from streams and ponds, as well as sub surface water from wells (both shallow and deep), and kuzhis (shallow pits).

During the 1960s itself, generous state subsidies were made available for the installation of pumping devices, at the rate of 25% for electric engines and 50% for diesel engines (Frankel 1972). As a result, in Palakkad district on the whole, the number of pump sets is reported to have risen from 79 in 1964–65 to 1592 in 1967–68 (a twenty fold increase within a three year period). In order to promote the lifting of water from smaller water sources such as ponds, wells and streams/rivers, a scheme such as the free supply of pump sets was introduced during the 1970s (SPB 1976: 1). Such a scheme was justified on grounds of increasing food production by assuring a third crop, as well as enhancing the utilisation of under-utilised water resources, implying that water sources such as ponds and wells were under-utilised. The State Planning Board also recommended the urgent energisation of pump sets, calling for special budgetary provisions by the State Electricity Board if necessary (ibid., 4).

As per a 2001 survey, there existed a total of 245 energised pump sets for irrigation purposes in the Kollengode and Elavenchery panchayats, with a horse power ranging from below 1.5 HP to above 5 HP.

TABLE 2.5 Number of Pump Sets in the Kollengode and Elavenchery panchayats

Panchayat	*Electric Pump Sets*	*Diesel Pump Sets*	*Total number of pump sets*
Kollengode	124	53	177
Elavenchery	60	8	68

Source: Kerala State Land Use Board 2001

Over the past ten years, an increasing number of tubewells have been dug, both to meet drinking water and irrigation needs. Borewells dug for irrigation is used to irrigate paddy lands as well as coconut plantations raised on parambu and paddy land.

The Introduction of New Seeds and Chemical Inputs

The introduction of high yielding varieties of paddy seeds and chemical fertilisers during the 1960s and 70s is reported to have led to a sudden spurt in yields in the area. Farmers however report that they were not able to sustain this increase in yield. The increasing unreliability in the supply of canal water since the 1980s is also reported to have decreased yields. An equally important factor according to farmers is the increasing pest attacks, which according to them has increased since the introduction of the high yielding varieties of seeds. While the yields differ in both seasons, the highest yields reported from the area average 3500–4000 kg per hectare. Farmers who suffer from unreliable supply of irrigation water (mostly the small farmers) record much lower yields, which can fall to a mere 1000 kg per hectare. While the high yields reported from the area are higher than the average yields for Palakkad district on the whole,[27] these are reported to be declining.

The introduction of chemical fertilisers has altered the traditional process of fertilisation of paddy fields in the region. Prior to the introduction of chemical fertilisers, manuring of paddy fields was elaborately undertaken during the months of February–March in order to meet the requirements of both the crops of paddy. Green manure was procured from forest patches near by as well as from the Tenmala, the hills that lay to the south. A wide variety of trees scattered across the plains, particularly in the upper slopes of the hillocks, in the parambu, and on pond bunds also provided much-valued mulch. Apart from *pachila valam* (manure from green leaves), *kali valam* (cow dung) was also an important source of fertiliser. This process of fertilisation was repeated more than once prior to sowing paddy around mid April. In addition, fields were sown with leguminous varieties such as horse gram and '*kollinnyal*' (a local leguminous variety), and on maturing they were ploughed back into the fields to enhance the nitrogen content of the soil. The application of green manures by growing and burying legumes in the soil is known to enhance the level of organic carbon and the availability of nitrogen, phosphate and potassium, along with other essential micronutrients. Even the incorporation of residues of legumes enhances the above-mentioned elements in the soil (Bhullar and Sidhu 2006).

Such practices are however on the wane. During the summer of 2002, I observed that very few fields were sown with *daincha*, yet another leguminous variety as a part of field fertilisation. Farmers observe that the increasing non-availability of organic manure and the ease with which chemical fertilisers can be applied has led to the increasing use of the same. Farmers however report that the continuous use of chemical fertilisers over the past three to four decades, and the corresponding decrease in the application of green manure has made the soil sandier in nature, with a reduced organic content. In an attempt to boost paddy yields, the government has devised schemes to address the problem. One such scheme implemented in certain parts of the watershed is the GALASA (Group Approach for Locally Adapted and Sustainable Agriculture Scheme) scheme, which encourages farmers to use organic fertilisers by providing it at subsidised rates.

Intensification of Paddy Cultivation: Implications on Water Consumption

The introduction of modern irrigation technology brought about fundamental changes to the traditional pattern detailed upon in the previous sections. A system that was based on delicately balanced moisture regulation and on highly restricted water use patterns was transformed into one where such upper limits were done away with. The disappearance of the traditional kalayi-potta classification for single and double cropping of paddy, of unrestrained lifting of water from ponds, changes in the timing of the agricultural calendar and the neglect of regular pond maintenance were some of the outcomes of this transformation.

Blurred boundaries between kalayi and potta

Converting single cropped lands into double cropped ones was the objective of most irrigation projects in the state. Referring to the proposal for the Vamanapuram Irrigation project in south Kerala, Santhakumar et al critique the irrigation department's urgency to convert the entire command area into triple cropped paddy land, irrespective of the existing wetland/dryland classification (Santhakumar et al 1995: A–33). The line of thinking during the late 1970s in the state for instance was to somehow enhance the extent of cultivable area, by converting dry lands into wetlands, and by increasing the extent of garden lands (Nalapat 1978). In the study area, as in many parts of Palakkad district, the introduction of canal irrigation led to an increase in the area under double crop (Santhakumar and Nair 1999). District level data indicates that the area sown more than once when expressed as a percentage of net area sown, increased from 31.7% in 1960–61 to 63.4% in 2004–05. More specifically, the percentage of area sown more than once increased from 31.7% in 1960–61 to almost double, i.e. 58.9% in 1975–76. This was the period when most of the irrigation projects were commissioned in the district.

TABLE 2.6 Changes in the area sown more than once as a percentage of net area sown in Palakkad district

	1960–61	*1975–76*	*1990–91*	*2004–05*
Area sown more than once as a percentage of net area sown	31.7	58.9	56.2	63.4

Source: Consolidated from Census Reports and Economic Reviews, 1976, 2005

As per the survey conducted by the Kerala State Land Use Board in the panchayats of Kollengode and Elavenchery in 2001, single cropped paddy land comprised a very small share of the total paddy land in the panchayat (0.63%) in the Kollengode Panchayat. The proportion of single cropped lands was higher in the Elavenchery panchayat (12.65%), which could be due to the increased unreliability of canal water supply here, (owing to its location in the tail end of the canal network of the Gayatri Irrigation Project), which acts as a deterrent in sowing a second crop of paddy particularly on potta lands.

TABLE 2.7 Paddy Land Classification in Kollengode and Elavenchery panchayats

Paddy Land	*Kollengode*		*Elavenchery*	
	Area in ha	*%*	*Area in ha*	*%*
Single cropped paddy land	11.5	0.63	198.25	12.65
Double cropped paddy land	1792.37	98.6	1368.5	87.25
Fallow	14	0.77	1.75	0.10
Total	1821.63	100	1568.5	100

Source: Kerala State Land Use Board 2001

Along with the external supply of water, the introduction of pumping devices also did away with the necessity of abiding by topographic constraints. In the earlier system, only the amount of water that flow out through the higher valve (mele ovu) was diverted to potta lands. Hence once the water level in the pond fell below the mele ovu, water had to be lifted out to potta lands were they in need of irrigation. Prior to the introduction of energised lifting, manual lifting was resorted to only when the need for irrigation was urgent. With the spread of double cropping, potta lands placed a heavy demand on the water stored in the ponds, and energised pumping was an enabling factor. In the absence of upper limits on the amount of water that could be so pumped to potta lands, the topographical distinction between kalayi and potta was done away with. Paddy cultivation was therefore no longer organised along drainage lines. Increasing reliance on water supplied through the canals however made the paddy crop increasingly vulnerable to variations in water supply as well. Water scarcity as observed today in the region, is mostly manifest in crop failure on potta lands. The kalayi lands are less vulnerable to crop failures.

The enhanced demand for water as a result of intensification of paddy cultivation has been compounded by the relatively greater water requirements of the high yielding paddy varieties in comparison to the older varieties. Farmers observe that while the new varieties boosted yields, they were unable to withstand water stress. When being transplanted, while the older varieties could be transplanted between 40–60 days of growth, the new varieties had to be transplanted by the 22nd day. If transplanting was delayed for want of water, it would reduce the yield farmers observe. In addition, chemical fertilisers had to be applied thrice before harvesting, for which availability of water was critical.

Replacing dry crops with wet, irrigated ones has been a phenomenon associated with the introduction of fairly regular irrigation facilities (Wilson 2002). With the spread of double cropping in the study area, the dry crops that used to be cultivated on potta lands were replaced with paddy. Farmers cite three reasons for this. One, the supply of water through the canals during the second crop season can damage a crop of beans or black gram, which do not require large amounts of water. Two, labour charges that farmers incur while harvesting a crop of black gram or groundnut will nullify any possible profit. The third reason they cite is the inferior status given to dry crops. Many farmers were very dismissive of dry crop cultivation on potta lands.

Sivan from Cheerani for instance had been getting very poor yields for paddy for want of adequate amounts of water during the second crop season. Yet, his response to replacing paddy with dry crops was '*Nyaan aa vaka jaathikal onnum cheyilya*' (I will not cultivate all those types). Many of the large farmers consider it shameful to cultivate dry crops when others cultivate paddy. Shankaran, a farmer owning 2.4 hectares of land in Manalipadam stated with great relief that to date he has been able to cultivate paddy during both the seasons, and has therefore been spared of facing public ridicule on account of raising dry crops. Others have not been so fortunate he admits, having to forego paddy cultivation fearing crop loss. Despite the enhanced water demands of paddy, the high prestige value associated with paddy prevents farmers from experimenting with dry crops.

Neglect of ponds

Velayudhan, a farmer from the area, reports that farmers in this area had heard of 'miracle stories' about the canals from farmers in the ayacut of other irrigation projects in the district, such as Malampuzha and Walayar. In anticipation of the water that the canals would bring into the area, a few ponds were converted into agricultural land. Some others were filled up soon after the canals were introduced. A total of 9 ponds[28] in the watershed are reported to have been filled up in such a manner. In addition, two ponds were filled up in order to facilitate the laying of the Left Bank Canal, for which the owners were given compensation.[29]

It needs to be noted however, that following the filling of a few ponds immediately before or after the introduction of the canals, only three have been subsequently filled up. Of these, two ponds in Velampotta were filled up by a landowner who had purchased the land here during the 1990s. He converted all the potta paddy fields into plantations of areca and coconut, and dug a tubewell to irrigate the same. He filled up the two ponds in anticipation of selling the filled up land as house plots. According to farmers in the area, the concerned landowner who was new to the area was aping the famous paddy land conversion and pond filling at Payilur Mukku in Elavenchery panchayat. Fertile double cropped paddy land and the Odakuzhi pond had been converted into housing sites at Payilur Mukku.

Other than the filling up of ponds, the introduction of the canals is also reported to have reduced the meticulousness with which ponds were filled with rainwater. Instead, the practice of filling ponds with canal water was initiated, a practice that continues amidst contestations and conflicts. Ponds were therefore not viewed as part of the catchment, but as mere storages of canal water, an aspect that shall be discussed in detail in Chapter 3. The secondary value attached to ponds was also manifest in reduced attention to regular pond maintenance. Regular desilting of pond beds, as well as packing of the bunds of the pond was gradually discontinued. This was also a result of changing land ownership patterns in the area, and issue that shall be dealt with in detail in Chapter 4. Very few ponds have been desilted in the post land reform era. If at all ponds are being desilted today, it is to meet the sand requirements of those who wish to convert paddy fields into house plots. In such cases the person in need hires an earth mover, commonly referred to as JCB, and digs out sand from the pond bed. The pond owner considers it beneficial, for his pond gets deepened without him incurring any cost. This was the case with the two cases of pond desilting reported in 2001–2002.

Box 2.2 Desilting the Pond with a JCB

In the desilting of the Poricholam pond undertaken by the owner Mohammed, the owner engaged a JCB and a tractor (the former to lift the sand and the latter to dump it on the fields of those who wanted it). He arrived at a deal with farmers who needed the sand, whereby the farmer paid the JCB owner Rs. 45 for each vetti (basket) of sand dug out and the owner of the tractor Rs. 55 per vetti (basket) for dumping it on his fields. In total the buyer had to pay Rs 110 per basket. Mohammed, the owner of the pond did not have to pay anything, but his pond was deepened in the process. A similar process was observed with regard to Panankavu Pond in Tahsildaar padam as well.

Source: Field Notes, 10th February 2002.

Photo 2.4 A silted up pond bed

The lack of regular desilting has resulted in many ponds getting silted up thereby reducing their storage capacity. Reduced storage capacity has implications not only for irrigation but also in facilitating drainage of water wwduring periods of heavy rainfall. When silted up, ponds are unable to swallow the enormous concentration of rainfall in a few days (Athreya et al. 1990), particularly the run off generated during the months between June–August when the monsoon is at is peak in the area. Athreya et al. argue that ponds cannot be merely seen as means of irrigation, but also as a means of drainage, and as such are important in preventing soil erosion and flooding. In addition they also impart stability to the system of well irrigation by recharging groundwater levels. Hence, from an ecological perspective, the decline of ponds is viewed as a threat to the entire ecosystem (ibid.).

Despite being given an 'underdog' treatment, ponds continue to play a critical role in sustaining paddy cultivation in the area. While they are unable to meet the entire water requirements of the high yielding paddy varieties, they provide critical interim relief particularly during the transplanting of the second crop, and while applying fertilisers and pesticides.

Box 2.3 Farmers' Views on Neglecting the Role of Ponds

Raman-'since the canals came, the care given to the ponds has reduced. It is similar to how the introduction of pipelines for distribution of drinking water led to the neglect of wells nearby'.

Muthukumaran -'After the canals came, '*naatile kulam arkum venda*' (no one wants the ponds in the village)'.

Farmers without access to ponds for instance are found to suffer from poor yields due to their inability to apply fertilisers on time (see Chapter 6). While the construction of ponds on a massive scale coupled with pumping facilities was suggested by some studies as a measure to reduce the risk to the winter crop (Frankel 1972), the reasons for the neglect of the existing ponds has not been considered worthy of attention.

Changes in the agricultural calendar: The breakdown of seasonality?

Following the development of irrigated agriculture, the intensification of paddy cultivation and in some cases the introduction of a second crop of rice has resulted in a breakdown of seasonality in the cultivation of rice in many rice growing regions of the world (Schneider 1995). As a result of the external supply of water, seasonality becomes less important than in rain fed systems (ibid.). In the study area, the introduction of the canal network brought with it a new schedule of water availability, which did not coincide with the monsoon cycle. Transplanting of the second crop, which would normally commence, by August–September has now been pushed to October–November. The primary reason cited by farmers is that the first release of water through the canal network arrives only by November. Most farmers therefore view raising seedbeds ahead of time as a risky proposition.[30] Many farmers, especially farmers in the tail ends of the second reach, wait for confirmation of water supply from the Chulliar reservoir before they commence cultivation, as the final release of water rarely moves beyond the second reach of the ayacut. Those farmers who would want to commence the second crop early with the aid of water stocked in the pond, are deterred by the fact that majority of the farmers would commence a month or more later. An early ripening of the crop would invite pest attacks.

The postponement in the commencement of the second crop has serious implications. For one it pushes the harvest into the month of February, and even March at times. Thus, the crop will have to face the brunt of the summer months marked by enhanced evapotranspiration rates, thereby aggravating the water requirement. In an evaluation of minor irrigation works in the state, the State Planning Board in 1975 recommended an advancement of the second crop through transplanting in order to make best use of water, as well as to minimise storage losses in summer when evaporation losses are highest (SPB 1975: 45).

The uncertainty in water availability at the time of transplantation leads many to sow according to the broadcasting method, referred to as '*chetu veda*' meaning sowing in slush. This method protects farmers from losses they sustain on expenditures for nurseries if water shortage at the time of transplantation or at the flowering stage causes damage to the crop. But the broadcasting method is more water consumptive. In the transplanting method the harvested fields are ploughed and transplanted with the water that remains in the field at the time of harvest, requiring little additional water. In contrast, while broadcasting seeds, the water is drained away from the harvested fields, following which intermittent irrigation and drainage is required to control the weeds. Farmers therefore report that while transplanting is more labour and cost intensive, broadcasting is more water demanding.

Using water sparingly

An interesting observation made by many farmers in the region is that the way in which farmers use water has changed since the introduction of modern irrigation technology. As Mohanan from Pulikalpotta remarked, 'The canals have made us callous while handling water'. While referring to the fact that farmers were less alert while opening and closing the outlets of the ponds, or while irrigating their fields, Selvan observed that in the past, not a drop of water was allowed to leak from the ponds. '*Ponninde podi pole nokum*' says Swami (meaning they would take care of water as they would take care of a particle of gold). According to Swami, today, when farmers spot a leaking outlet, they may not act as hastily as before, primarily because they feel that the canals will supply water again twenty days later, and they could re fill the ponds. Hence a small leak is not taken seriously. Others like Sivan attributes this laxity to the impression that 'this is canal water that the government supplies, '*enkineyo pote*' (let it go anyhow). Farmers feel that this was not so when the water in the pond was limited in availability. Another example they cite in this regard is that the present practice is to fill the fields with water. In the past, much lesser amounts of water was let into the fields. '*Nanachu, nanchu vidum*' (meaning that they would only wet the crop, not flood it with water). Today, when water runs through the canals, farmers will take as much as possible, not as much as needed, they say. The possibility to lift water easily out of wells and streams has also led to this change in approach farmers observe.

Conclusion

Two issues have been raised in this chapter. The first concerns the desirability of a paddy centred cropping pattern, and the second is the suitability of a single model of irrigation that is based on the construction of reservoirs and canals.

Kendy et al argue that a critical assessment of the hydrological impact of agricultural policies help to understand the underlying causes of water scarcity in food growing regions (Kendy et al. 2003). While the cultivation of paddy is critical in ensuring food security in the state, the hydrological implications of such a paddy centred agricultural policy have been overlooked. Intensification of a water-consuming crop like paddy can place enormous pressure on existing water supplies, particularly in regions such as the Palakkad Gap. One such hydrological impact was the abandoning of water saving strategies (like restriction of double cropping and the synchronization of the agricultural calendar with the monsoon cycle) following the conversion of single cropped lands into double cropped ones, which has been induced by the provision of canal irrigation. The other is the gradual disappearance of dry crop cultivation on single cropped land, to the extent that farmers feel that they would suffer from a loss of prestige if they were to cultivate a dry crop like black gram in the place of paddy.

It is interesting to note that the paddy focused agricultural policies of the state had been critiqued by the government as early as 1975. The State Planning Board for instance had critiqued the practice of advocating a third crop of paddy on grounds of non availability of adequate amounts of water, and instead advocated the cultivation of pulses, sesamum and horsegram (SPB 1975:45). In the case of Palakkad district for instance, even a second crop of paddy was not recommended on grounds of its low water-use efficiency (KAU 1991:120 in Santhakumar et al. 1995:A–34). At the policy level however, there has been no explicit focus on encouraging the cultivation of water efficient crops listed above. Amongst food crops, the existing incentive regime is still in favour of paddy, with paddy farmers being given a bonus at the end of each cropping season along with subsidized supply of chemical fertilisers. In 1998, after about thirty years of the functioning of the Gayatri Irrigation Project, during which period crop failures on potta lands were not uncommon, the focus continued to be on the intensification of paddy cultivation by encouraging double cropping (GOK 1998). The intriguing fact is that despite the decline in area under paddy cultivation in the state, the government continued to focus on designing major irrigation projects that would enhance the cultivation of paddy. The absence of a policy environment that emphasised and promoted the cultivation of water conserving food crops, has led to a situation wherein paddy is being replaced with crops such as coconut, or banana, neither of which are water conserving. This has been the case with Palakkad district as well.

The second issue concerns the propagation of a single model of irrigation throughout the state, and its inability to achieve the objective of increased paddy production. Such a model of irrigation has been critiqued for paying scant attention to area specific agricultural and water management practices (Santhakumar et al. 1995). This has been illustrated in the manner in which modern canal irrigation was introduced into the study area. The new schedules of water delivery through the canal network and the introduction of energized pumping led to a situation wherein the local topographic and climatic features began to matter less. As one farmer put it, when water began to be available in plenty through the canals, why bother about kalayi and potta? This single model of irrigation has also been critiqued for its disregard of locally available water supplies, of the relevance of minor storages and in situ moisture conservation, all of which are relevant in an undulating topography that is typical of most parts of the state. During the past two decades, the value of rainwater harvesting and in situ moisture conservation has been increasingly emphasized in policy documents (GOK 2004), particularly in the context of watershed regeneration programmes. Interestingly these elements were reflected in the ponds of the study area. They were small, dispersed storages of water that collected locally generated run off, and in the process enhanced in situ moisture conservation.

In short, an agricultural and irrigation policy that is specific to the agro climatic specificities of the area would give more emphasis on irrigation cum water harvesting structures like the ponds. This is particularly so as the irrigation potential of canals has been on the decline in the district. Over the last five years, the net area irrigated by government canals has decreased by 15% in the district. The inadequacy in water supply through the canals coupled with the declining storage capacity of ponds has however not prompted a reconsideration of the presently followed water intensive cropping pattern. Neither has it prompted a reconsideration of the existing mode of irrigation. To the contrary, it has led to a more intensive exploitation of existing water sources, manifest in the increasing number of tubewells and lift irrigation schemes, the latter drawing water from the declining flow in rivers and streams.

Notes

1 'The success following the Second World War of plant-breeding programmes aimed at increasing crop yields in the agriculturally underdeveloped tropical areas of the world gave rise to the term *Green Revolution*. The aim of the programme was to produce varieties – particularly of wheat and rice-which would be capable of high yields: the HYVs as they became known. . . . The principal outcome of the Green Revolution was to produce dwarf or semi-dwarf varieties of cereal crops with stiff stems and short upright leaves which allowed dense planting, with minimum shading and relatively restricted root systems, and the potential to give high yields when supplied with adequate fertilizers, water, and disease protection' (Tivy 1990: 106).

2 Shah (1993: A–74) delineates two phases in Indian agricultural policy and planning in the post Independence era. The pre 1965 era which was dominated by a mix of Nehruvian, Gandhian and Marxian visions, where the national policy stressed on land reforms, community development and cooperative farming. The approach in the post 1965 era according to Shah emerged out of a disillusionment with the above approach and was also in response to the dire need to increase food production. It was also spurred by the availability of high yielding varieties of Mexican wheat and maize varieties.

3 Development planning in India in the post independence era (i.e. since 1947) has been executed through the formulation and implementation of the five-year plans. The Planning Commission of the country has been entrusted with the responsibility of developing, executing and monitoring of these five-year plans. The first five-year plan commenced in 1951, and India is currently in its eleventh five-year plan (2007–2012).

4 The physical area under paddy cultivation irrespective of the number of times the crop is raised during the year is termed as the net area under paddy. Gross area under paddy is obtained by adding the area under paddy crop as many times as the crop is harvested in the year or in other words it is the total crop area under cultivation during autumn, winter and summer of the year (Kerala Statistical Institute 1994: 6).

5 It is pertinent to note that the conventional view of food security as increased food grain production led to policies that were exclusively focused on paddy, at the cost of coarse grains, and a variety of tubers and fruits such as jackfruit, mangoes and papaya, which were equally important dietary supplements for the average Malayalee. Particularly so

as different parts of the state exhibited significant diversity with respect to rainfall, terrain and vegetation patterns which thereby supporting a diverse array of food crops, including vegetables and fruits.

6 The Intensive Agricultural Development Programme (I.A.D.P.) was a pilot programme undertaken in 15 districts of the country of which Palakkad was one. This programme was initiated in 1961, and it emphasized the necessity of providing the cultivator with a complete 'package of practices' in order to increase yields, including credit, modern inputs, price incentives, marketing facilities, and technical advice (Frankel 1972).

7 The physical area under paddy cultivation irrespective of the number of times the crop is raised during the year is termed as the net area under paddy. Gross area under paddy is obtained by adding the area under paddy crop as many times as the crop is harvested in the year or in other words it is the total crop area under cultivation during autumn, winter and summer of the year (Kerala Statistical Institute 1994: 6).

8 The most recent report of the Government Rice Commission (1999) was the outcome of the thirteenth government level enquiry into this issue.

9 Analysing the impact of irrigation, rainfall and fertilizer use per hectare on agricultural productivity, Kannan and Pushpangadhan conclude that the effect of irrigation as calculated using the water availability index was not significant as far as productivity was concerned (Kannan and Pushpangadhan 1988).

10 As of 2004–05, average paddy yields for the state was estimated at 3582 kg/ha (GOK 2006b).

11 Unni reported that while the net area under paddy has been falling in certain districts, including Palakkad, coconut has been gaining. The districts (particularly the northern districts of the state) that showed the greatest tendency to shift away from rice also experienced the greatest increase in the area under coconut cultivation (Unni, 1983).

12 Wages of hired agricultural labour comprises the most important item of operational cost in the cultivation of paddy in the state. The high labour input required in paddy cultivation especially during transplanting, periodic weeding and harvesting takes a heavy toll.

13 These include the Malampuzha, Pothundy, Gayatripuzha, Walayar, Mangalam, Chitturpuzha, and the Kanjirapuzha irrigation projects.

14 During the period 1995–96 to 2004–05, Palakkad registered a 20% decrease in area under autumn paddy as compared to the state average of 44%, and a 11% decrease in winter paddy as against the

state average of 34%. In the case of summer paddy however, it has registered a substantial decrease of 89% of summer paddy as against the state average of 40%. With regard to production, Palakkad registered a decline of 8%, 1% and 93% for the three seasons for the above period as compared to the state average of 30%, 27% and 40% (GOK 2006b).

15 In 2005–06, area under banana cultivation in Palakkad was 11248 ha, next only to Wayanad district where 12842 ha were put to banana cultivation (GOK 2006a).

16 This is computed by comparing the targeted net area to be irrigated with the area actually irrigated (Vishwanathan 2002). While the net area targeted to be irrigated under the Gayatri Project was 7651 ha, the area actually irrigated was only 4880 ha. (Vishwanathan 2002, on the basis of summarisation from various issues of the Economic Reviews) The utilisation figures for the other irrigation projects in Palakkad district were 67%, 53%, 83%, 69% and 103% for the Malampuzha, Pothundy, Walayar, Mangalam and Chitturpuzha projects respectively.

17 One of the conclusions of the government's Expert Committee constituted to study the problems facing the paddy sector was that 'paddy lands are not distinct and independent entities, even where paddy is the major crop, but part of mutually dependent resource continuum' (Govt. Rice Commission 1999:110).

18 Francis Buchanan, deputed by the British government to survey the region in 1800 (see p. 30–31 in Chapter 1), cites a similar classification, using a different terminology though. Rice land as observed by Buchanan was classified into two kinds, Palealil lands which were the 'higher parts of the rice-ground, and never produce more than one crop in the year' (Buchanan 1870: 70) and 'lower parts of the rice-land' termed Ubayum, a great majority of which produce two crops annually (p. 71).

19 A territorial division of olden times.

20 *Varagu, ragi* and *chama* are different kinds of millets. While varagu refers to proso millet, ragi refers to finger millet and chama to little millet. *Cholam* refers to corn.

21 Sengupta for instance reports a similar pattern from parts of Orissa in eastern India, wherein the cropping pattern adopted by the indigenous people had an in built mechanism of drought resistance. Sengupta observes that while short duration rice, requiring less moisture which was ready for harvest within 40–60 days of sowing was planted on the '*at*' or the uplands, long duration rice was planted in the comparatively secure lowlands or '*bahal*' to support a longer growth period of

120–160 days. Dry land rice in combination with coarse grains, pulses and oil seeds ensured optimum food supply even during periods of erratic rainfall. Vegetables were grown on homestead land, and fruits and tubers from the forest supplemented the diet (Sengupta 2000).

22 'Water flowing from the forest to the tank and then to the paddy field, contained forest-soil runoff that was rich in elements of fertility. This water also brought silt in suspension and other plant nutrients in solution, renewing the soil fertility at least to some extent every year. . . . The nutrients and silt accumulated in the tank were transported by water-flow to the paddy fields when the mud sluice or *mada horrowa* (the sluice in the lowest situation) was opened and the tank was de-silted and its contents were added to the paddy tract, increasing its fertility'. (Ulluwishewa 1991:105). In a similar description, Todd in an account of a farm in Central Java describes how water entering the farm via an aqueduct or ditch along the contour of the farm was charged with nutrients so that it could fertilise and irrigate the crops (Todd 1984:154).

23 Vishu is the astronomical New Year day festival celebrated on April 14th in the state of Kerala, coinciding with the first day of the Malayalam month of *Medam*. This occasion signifies the Sun's transit to the zodiac Mesha (*Mesha Raasi*), which is the first zodiac sign as per Indian astrological calculations. Similar festivals are celebrated around the same time in the states of Tamil Nadu, Bihar, Assam and Punjab. In Kerala, Vishu marks the onset of the new agricultural cycle.

24 Buchanan too reports of such a practice, noting that ' The *Maytam,* or ground kept for raising seedlings, is chosen in a high situation and poor soil. It pays neither rent nor land tax' (Buchanan 1870: 71). From the description it seems that *maytam* refers to naatupottas.

25 Poochamban, Mattachamban, Navara, Ongu Modan, Vellachamban, Tavalakannan, Cheera, Karipali were some of the traditional varieties used during the autumn crop.

26 Pandian while discussing the process of agrarian change in Nanchilnadu in Tamil Nadu, describes changes brought about to the 'irrigation milieu' with the introduction of modern irrigation technology through the introduction of a reservoir based canal network into the region. I use the term agricultural milieu to refer to the changes that have been brought about to both agricultural and water management.

27 During the autumn and winter paddy crops for the years 1999–2000, 2000–2001 and 2001–2002, the average paddy yield for the district

on the whole was estimated at 3412.5 kg/ha, 3256.5 kg/ha and 3492 kg/ha (GOK 2006b).

28 These include Nellikuzhi and Panneri in the Manalipadam padashekhara samity, Odakuzhi in Payilur Mukku, a *chera* in Velampotta and another in Kallatukolambu, a pond in Cheerani, two ponds in Kollengode (one in front of the Kollengode Urban cooperative bank and the other behind the mosque), and Kannankulam in Kizakemuri.

29 These include the potta near Tayankulam and a pond in Mannathupara. Sankaranaryanan in Mannathupara, whose lands lie close to the left bank canal lost a quarter of his pond in such a manner, for which he did not receive any compensation.

30 The practice of raising seedbeds on the *naatu potta*s has also been abandoned as most of the high land reserved for *naatu potta*s has been converted into paddy fields, coconut plantations and even housing sites.

3

Water for Irrigation

The Missing Ecological Dimension

'Water's seeming ubiquity has blinded society to the need to manage it sustainably and to adapt to the limits of a fixed supply'.
Postel 1984: 18.

'The supply-side approach has quite unceremoniously careened off the bend and noisily crashed against ecological limits'.
D'Souza 2003.

This chapter describes the evolution of large-scale infrastructure for irrigation in the study area, primarily through the development of the Gayatripuzha Irrigation Project. It describes the long drawn process of water transfer from one basin to another through dams and transfer canals, until it reaches the Chulliar reservoir. It also highlights the ecological dimensions that have been neglected in the process. The chapter also looks into how the predominance of this infrastructure oriented approach leads to a conceptualisation of water scarcity in supply terms alone, as a result of which ameliorative measures tend to be focused on further infrastructure creation. The chapter also describes how the introduction of the large-scale canal system resulted in an undervaluation of the existing local water resources in the study area, viz. the ponds and the streams. Presently, the ponds and the streams in the region function as mere appendages to the canal system, a phenomenon that has important implications on local water availability. Finally, the chapter discusses the relevance of local level water resources planning and development in dealing with the water crises of the region. This is an area that has received much emphasis since the implementation of decentralisation in the state since the late 1990s.

Water from Afar: The Long Journey through the Canals

This section details the prevailing irrigation model in the district of Palakkad and the study area, viz. one that is centred around the construction of medium and large reservoirs and canal networks. As mentioned in Chapter 1, the Bharathapuzha basin in which the study area is located, contains ten reservoirs intended for irrigation, of which seven are located within the boundaries of the state of Kerala.[1] The present section focuses on the functioning of the Gayatripuzha Irrigation project located in the Bharathapuzha basin, which supplies water to the study area.

As mentioned in Chapter 2, water began to be supplied through the canals to the panchayats of Kollengode and Elavenchery by 1970 with the completion of the Chulliar reservoir. The Chulliar dam is constructed across the Chulliar River, a tributary to the Gayatri River (see Figure 3.1). The Chulliar dam along with the Meenkara dam (constructed across the Meenkara River, yet another tributary to the Gayatri River) comprised Stage I and Stage II of the Gayatripuzha Irrigation Project. While both the dams together are expected to irrigate a command area of 5463 ha, the Chulliar dam alone is expected to irrigate a command area of 3074 ha (GOK 1998: 33). The Chulliar reservoir with a storage capacity of 13.7 million cubic metres has a catchment area of only 27.80 sq.kms. (GOK 1998).[2] This comparatively small catchment area of the Chulliar has lost most of its forest cover. The resultant soil erosion of the catchment slopes has led to siltation of the reservoir bed,[3] which further reduces the storage capacity of the reservoir. Reduced storage capacity, coupled with the unequal distribution of water between the head and tail reaches of both the main and distributary canal systems has resulted in a situation wherein the existing storage, without an additional supply of water is unable to meet the water requirements of the entire command area.

FIGURE 3.1 Details of the transfers of water from the Chitturpuzha (Aliyar) basin to the Gayatripuzha basin, cutting across the inter-state boundary

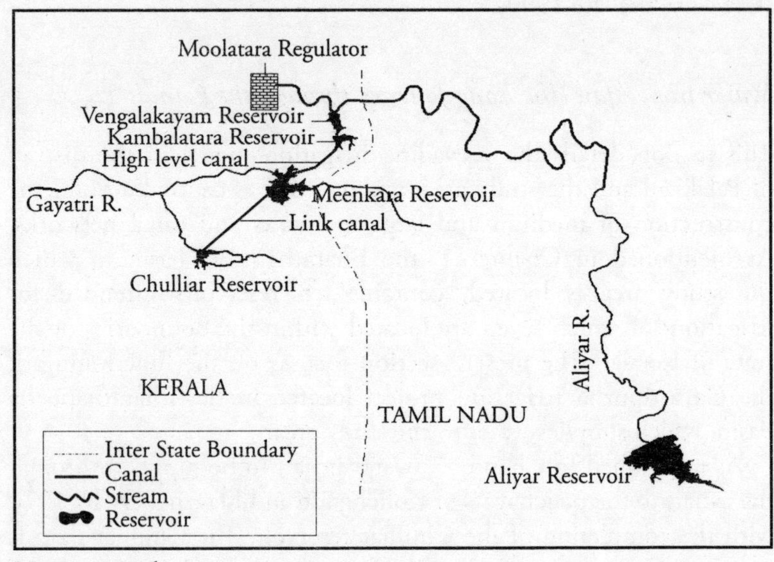

Map not to scale

Water from across the state

Inadequate storage in the Chulliar reservoir has been an issue ever since the inception of the dam in the late 1960s. A well to do farmer from the area recalled that at the evening party that followed the inauguration of the Chulliar project, attended by engineers from the irrigation department, politicians and other influential members of society, the engineers remarked that the storage in the Chulliar would be insufficient for meeting the water needs of the entire command area, and that an external supply was required. It is not clear whether diverting water from the adjoining Chitturpuzha irrigation system was conceived of at that time, but by 1982 a 4.2 km link canal was constructed, carrying so called surplus waters from the Meenkara reservoir to the Chulliar reservoir. This water was originally transferred to the Meenkara reservoir from the adjoining Chitturpuzha (Aliyar) system. It is to be noted that the Chitturpuzha River is known as the Aliyar River upstream.

The transfer of this 'surplus' water from the Chitturpuzha system to the Gayatripuzha system is not a simple one, but part of an inter-state, inter-basin agreement to share the waters of three rivers that flow through the states of Kerala and Tamil Nadu. As per the Parambikulam-Aliyar Inter-State Water Sharing Agreement[4] (commonly referred to as the Parambikulam Aliyar Project or PAP), the two states agreed to share the waters of the rivers Periyar, Chalakudypuzha and the Bharathapuzha, all of which originate in Tamil Nadu, but later flow into Kerala. Despite only a small proportion of the overall catchment area lying in the state of Tamil Nadu, the agreement led to substantial diversion of the river waters into Tamil Nadu through the construction of a series of ten dams, reservoirs, tunnels and canal systems in the upper catchments of these rivers (Ravi et al. 2004).

The Chitturpuzha and the Gayatripuzha are located in the basin of the Bharathapuzha, one of the three rivers whose waters have been subject to the above inter-state division. As a result of the inter-state division, water that flowed down through the Chitturpuzha into Kerala was now held back behind three dams, viz. the Aliyar, Upper Aliyar and the Tirumoorty dams. This water was then diverted across basins into the state of Tamil Nadu. Keeping in mind the water requirements of double crop wetlands in the Chitturpuzha valley, the state of Tamil Nadu as per the agreement promised to release to the state of Kerala, 7.25 tmc ft (thousand million cubic feet) of water annually from the Aliyar dam into the Chitturpuzha basin at the Manakadavu weir in Kerala close to the inter-state border (GOK 1998). It is a part of this annual release along with the floodwaters of the Chitturpuzha system during the monsoons that makes its way to the Chulliar reservoir. The transfer of water from the Chitturpuzha scheme to the Gayatripuzha scheme is also necessitated by the reduced storage capacity of the reservoirs in the former. The former is essentially a diversion scheme, with two small storage reservoirs (the Venkalakayam and Kambalathara Eris) that are inter-connected (see Figure 3.1)[5]. The water in excess of the requirements of the Chitturpuzha system is released through the Kannimari surplus escape into the Meenkara reservoir in the Gayatripuzha system. This surplus escape takes off from the Left Branch Main Canal of the Chitturpuzha system.

Once water is stored in the Meenkara reservoir, the 4.2 km high level link canal is opened through which water is taken to the Chulliar reservoir. This marks the last phase in the long journey of water from the Aliyar reservoir to the Chulliar. This journey is marred by intense contestations; it has been particularly so during the past seven years (1995–2002). The farmers on the Kerala side allege that Tamil Nadu has been releasing less than the promised amount,[6] due to which the farmers in the Chitturpuzha ayacut are less inclined to release any water to the Gayatri system. Farmers from the ayacuts of the Meenkara and the Chulliar hire jeeps and go to the Moolatara regulator in the Chitturpuzha system from where water is diverted to the Gayatri system. They persuade the *maestry* (or the *lascar* as he is referred to) stationed there (with money and liquor) to release the water through the Kannimari surplus escape into the Meenkara dam. Off late, when the farmers from the Chitturpuzha ayacut have not been getting adequate amounts of water, they have opposed the diversion of water into the Meenkara system. If the water does reach the Meenkara dam, farmers from the Chulliar have to ensure that the maestry at the dam site opens the link canal to Chulliar. Once again, the farmers from the Meenkara command area resist this move as it reduces the water available to them. At each point therefore diversion is contested. The Chulliar dam therefore is located at the tail end of a series of surplus escapes and link canals, making it extremely vulnerable to inadequate and unreliable supply.

The long-term ecological implications

Diminishing flows in the Bharathapuzha River have been linked to the damming of almost all its feeder rivers (Prabhakaran 2003:60) It has been established that where a number of dams have been sited on a river, the cumulative impacts on downstream flows, water quality, natural flooding, and species composition are complex, leading to considerable losses, much of which could remain undetected for long periods of time (Parasuraman and Sengupta 2001). The specific impact of the diverting of the Chitturpuzha waters from the Chitturpuzha and thereby the Bharathapuzha basin on downstream flows has not been assessed. It has been critiqued that while formulating the inter-state treaty, meeting the irrigation demands of Chittur taluk was the only point of consideration, due to which 7.25 tmc ft was promised to be allocated. The impact of upstream diversions on downstream flows in the Chitturpuzha and the Bharathapuzha rivers was ignored (Prabhakaran 2004).

Another ecological impact of large multi basin transfers like the PAP is the dam-induced degradation of river catchments. While dam induced submergence of forests in the upper catchments is one factor, the opening up of hitherto dense forest pockets and their resulting degradation is often overlooked. Nair observes that the construction of dams in the upper catchments of the three rives involved in the inter-basin transfer of water has resulted in deforestation, conversion of natural forests into natural plantations and the conversion of contiguous extensive forest tracts to fragmented forest pockets leading to further degradation (S. C. Nair 1991). Degradation of river catchments is not confined to the reservoirs constructed as part of the PAP agreement alone. None of the dam catchments in the Bharathapuzha basin in Palakkad deliver a perennial stream (S. C. Nair 2004).

Siltation of reservoirs has been the natural outcome of deforestation of the catchments of the reservoirs, aggravated by the rainfall pattern in the state where in a substantial amount of the annual rainfall is concentrated during a span of four to five months. Apart from deforestation, increasing encroachments into forestland have aggravated the problem. It is to be noted that the state policy during the 1960s and 70s encouraged such private encroachments in order to make available land for agriculture (Narayana and Nair 1983). While it has been acknowledged that inappropriate land use and forest management practices in the catchments result in siltation (Santhakumar and Rajagopalan 1993), efforts to redress siltation however remain focused on structural solutions such as the creation of dead storages to store the silt (Santhakumar and Rajagopalan 1993). It has not led to efforts to rehabilitate catchment forests in the existing dams, which would augment the lean season flows into the reservoir (S. C. Nair 2004).

The Palakapandi solution to water scarcity

A cost benefit analysis of most dam projects planned and implemented in the early dam building era, would reveal that environmental values have not been considered, primarily because they did not enter into the equation between costs and benefits (Postel 1984: 31). The PAP project was formulated at a time when water and energy were considered to be available in plenty in the state of Kerala. During the 1950s and 1960s for instance, Kerala was considered to possess 'surplus' water, which could be liberally diverted to 'needy' states like Tamil Nadu (George and Krishnan 2000).

It is intriguing however that even at present when the declining availability of fresh water has been acknowledged, the solutions proposed to ameliorate the scarcity situation, take no cognisance of such environmental values. The proposed Palakapandi and Kuriarkutty-Karapara projects are examples. The proposed Palakkapandi diversion scheme (currently under implementation) proposes to construct a long tunnel through the steep slopes of the Tenmala that drain into plains. This is in order to divert the peak monsoon run off in the Palakkapandi stream into the Chulliar reservoir. The impact of this diversion on downstream flows as well as on the numerous lift irrigation schemes that draw their supply from the Palakkapandi stream has been overlooked. The justification given is that it is only the monsoon run off that would other wise run down wastefully, that is diverted into the Chulliar reservoir. The Kuriarkutty-Karapara project is a multi-purpose medium irrigation project that envisages the construction of a storage dam across the Karapara River and a diversion weir across the Kuriarkutty (GOK 2002b). Both these rivers are located in the adjoining Chalakudy River basin, and are tributaries to the Chalakudy River. The water to be supplied by this scheme is proposed to be made available to the Chitturpuzha system and thereby to the Chulliar system as well.

Work on the project has however been stalled owing to environmental objections raised against its location in fragile forest tracts.

Managing Local Water Resources: The Case of the Streams and the Ponds

This section focuses on the management of local water resources, namely the streams and the ponds in the Vararyiri watershed. These were the two primary sources of irrigation for paddy cultivation prior to the introduction of the canal network. As discussed in Chapter 2, the introduction of modern irrigation (through the canal network as well as through the introduction of energised pumping) has brought about significant changes in the management of ponds. I argue in the following sections how the introduction of modern irrigation has led to an under valuation of the irrigation potential and the conservation value of these local irrigation sources.

Ponds: from water harvesting to containers of canal water

To briefly recapitulate the discussion on ponds in Chapter 2, ponds in the study area were not pure irrigation structures. An important function of ponds was the intermittent storing of run off and enriching the water table through the percolation of standing water. Such a dispersed system of water harvesting helps to store some of the run off that is generated during the south-west and north-east monsoons put together, which accounts for more than 90% of the annual rainfall in the area (see Table 2.4 in Chapter 2).

The slowing down of run off helped in reducing the impact of flooding downstream during periods of heavy downpour as well. It also helps in checking soil erosion, particularly so as the uplands have lost most of their tree cover. In addition, the positioning of ponds and double cropped lands along the drainage lines in each micro catchment enabled a system of moisture regulation to evolve that was critical to paddy cultivation. When viewed cumulatively, the 150 odd ponds in the 1450-hectare Varayiri watershed, slowed down run off, which was then made use for agriculture and drinking water (by recharging near by wells and pits).

The intervention of the canal system in the local hydrology marks an important shift in the management of water resources in the pond systems. For one, they have at certain points disrupted the flow of natural drainage into ponds. This is particularly so in the case of the main canal, which traverses through the highest ridge line of the watershed. At Mannathupara for instance, drainage from the rocky patches, which earlier drained into the ponds in the higher reaches, began to flow down through the main canal. Kannakulam in the Kizakemuri padashekharam, located below the main canal, began to receive lesser run off after the construction of the main canal. In another instance in Mannathupara, a farmer was compelled to convert his pond into agricultural land, as the pond never filled up after the laying of the canal. Surface run off flowing down the canals has reduced inflow not only into all the ponds located below the main canal, but also into ponds that are located below distributary and field channels that traverse through sloping terrain. In some cases, overhead spouts have been placed, but they transfer a very limited amount of the run off into ponds. This issue has not been given much thought by the Irrigation Department, presumably because ponds were not ascribed with significant irrigation potential.

It was also observed that after a heavy spell of rain, mounds of sand would be deposited on the bed of the main canal and the bed of field channels that are constructed at the lowest end of a slope. The field channels on the lower slopes of the Mookarshan hillock and those at Peringotukavu are examples of the same. This indicates that significant amount of run off flows through the canal, which would have normally drained into the pond below.

Another factor that contributes to reduced surface inflow into ponds is that farmers do not take as much effort as before in diverting water into ponds. Cleaning of drainage channels bringing in surface run off into ponds is no longer done with the same vigour, according to farmers. It needs to be noted that the cleaning of these drainage channels was undertaken by farmers from each pond ayacut and did not involve any intervention from the irrigation department. Compared to the past when they would divert the run off generated from even the intermittent showers into ponds, today very few farmers take the effort to collect even the substantial run off generated during heavy spells of rain. In contrast to the spread out rainfall pattern, canal water is released only during two months in a whole year. Hence, while directing run off throughout the year into ponds sustained a minimum supply of water into the ponds for a longer period of time, directing canal water into the pond ensures supply only for a limited period of time.

A more significant outcome following the introduction of canals has been the filling up of ponds with water supplied through the canals. In some cases, the outlets along the field channels of the canal system have been conveniently positioned so as to easily divert water from the canals into the ponds.[7] Even where outlets were not placed immediately above the ponds, minor undulations in the landscape and the terraced paddy fields enable farmers to manoeuvre the flow of canal water into the ponds. The filling of privately owned ponds with public canal water that flows through the canals has important implications in terms of equitable distribution of the existing water supplies, an issue that shall be dealt with while discussing water distribution in Chapter 6.

For the present discussion it is important to assess the ecological dimensions of this phenomenon. According to Raman, a small farmer from Manalipadam,

> Just as one fills up vessels at home with drinking water during the summer months (to store as much as possible), so also one has to fill up ponds with canal water, as run off and rainfall alone do not fill them up as before.

The most significant outcome of this shift in approach is that ponds are less viewed as catchment based storages of internally generated run off and more as containers of externally supplied canal water. Such a view is endorsed by the irrigation department as well as in policy documents. The irrigation department's proposal to enhance the water use efficiency of the Gayatri Irrigation project for instance makes note of the need to augment the storage in ponds by filling it with canal water (GOK 1998). Similarly, a scheme for the free supply of pump sets by the state government during the 1970s was justified on grounds of enhancing the utilisation of 'under-utilised' water sources such as ponds, wells and streams (SPB 1976:5). The underlying assumption was that the above-mentioned water sources were under utilised for want of lifting devices. Ponds were therefore viewed only as pumping sources for irrigation, their value as catchment based water-harvesting structures being overlooked. Since the pond is no longer defined in terms of its catchment and command, management of the catchment in a way that maximises inflows into the pond is no longer considered to be a necessary part of pond management.

The Varayiri – a dying stream

A third source of irrigation, particularly for farmers in the lower reaches of the Varayiri watershed, is the Varayiri stream. Smaller streams from the higher reaches of the watershed, such as the Maripadam todu (todu being the Malayalam word for stream), the Vellanara todu, the Pulikalpotta todu and other smaller streams pool their flows in the Varayiri todu. Farmers unanimously report that flows in all the smaller streams, as well as in the Varayiri stream have significantly declined over the past 30–40 years. Farmers recall that even twenty five to thirty years ago, smaller streams such as the Maripadam used to flow for most of the year, except during the peak summer months. Today, these streams flow only during the monsoon months and for a few weeks after water is supplied through the canal system. This is true for the Varayiri stream as well, which is no longer perennial. It is more of a 'conduit of rainwater'[8] and canal water that has not been soaked up by the land in its catchment. The absence of any flow data in this regard makes it difficult to substantiate this point with evidence other than farmer's observations.

Declining flows in the stream warrants a closer look at the catchment. Studies have well established that stream flows are significantly related to land use changes in the catchment (Calder 1998). Catchment studies have demonstrated the water conserving function of natural forests, with indications that increased infiltration under natural forests lead to higher soil recharge and increased dry season flows (ibid.)[9]. Degradation of the upper catchments of streams and rivers has been shown to increase overland runoff resulting in spate flows. Changes in land use patterns can lead to reduced infiltration capacity of the soil resulting in peak flows during the rainy season (Bandyopadhyay 1987). Decreased infiltration also fails to recharge underground storage aquifers, in turn failing to maintain springs and to supply dry-weather stream flow. (Bruijnzeel 1990, Pereira 1973: 79).

The Varayiri watershed has undergone significant land use changes during the past three to four decades. The most significant change has been the clearing of forest vegetation (see Chapter 1 for a description of the forest cover that used to exist in the surrounding regions) and the intensification of agriculture. Farmers in this area recall that prior to the 1970s, the rocky, uncultivated areas were covered with fairly thick tree cover. Krishnankutty from the Kannankolambu area, whose fields are located immediately below the main canal that traverses through this area, recalls that the trees on the uplands above the main canal were cleared by landowners (referred to as *janmis*) prior to the nationalisation of private forests in 1971 (see Chapter 4). His elder brother Vasudevan remembers lighting fires to scare away tigers. Further down along the main canal, Ramachandra Iyer who owns land in Cheerani, recalls that there used to be good teak forests in the area around the Dharmi Temple, frequented by wild animals. These animals would come right up to their backyard, while preying on cattle. On losing their vegetative cover, small hillocks have been reduced to exposed rocky patches, which are now being increasingly converted into rock quarries. Sixty-seven hectares in the Kollengode panchayat comprise of such rocky areas (1.61% of the total land area in the panchayat) (Kerala State Land Use Board 2001).

Though agricultural lands predominate in the watershed, they were not devoid of vegetation cover. The small hillocks, dispersed throughout the watershed, that slope down into valleys were covered with a variety of trees. Farmers also recall specific trees that grew on pond bunds, as well as on bunds of paddy fields. Along with the nationalisation of private forests, the emerging settlement pattern has also displaced the vegetation cover.

Box 3.1 Deforestation of the Cheerani Hill

We were spared from the summer sun as it was a cloudy day. We decided to climb the Cheerani hill (163 msl), the highest point in the watershed. So small a hill compared to the Tenmalas in the background. We walked past Mukilele *Potta* (referring to the pond located at the highest level), which collected run off from the eastern slope of the Cheerani Hill. We then walked through a colony of houses (belonging to agricultural labourers) located on the lower slopes of the hill, the land for the same having been surrendered by Kailas Iyer, an erstwhile landlord of the area during the land reforms of 1970. The Cheerani Hill was also under his ownership, which was vested with the Forest Department, following the nationalisation of private forests. As we climbed the hill, we could see the main canal to its south, and the adjacent watershed, which drains into the Vazhapuzha stream. Abu Bakr, a seventy-year-old farmer in the area remembers the large trees that grew on the hill, which were felled about forty years ago. The hill today is severely degraded. With all the soil washed away, it resembles a heap of rocks. And in place of the large trees that Abu Bakr narrates of, there remains only thorny shrubs and dry grass.

As we returned to Nenmeni, it began to rain, the first of the pre monsoon showers.

Source: Field Notes, 4.5.2002

Most of the housing clusters in the watershed are located on sloping, higher ground including parambu land. As a result, most of these clusters are located in the catchment of ponds (see Figure 2.2 in Chapter 2). The entrenchment of the nuclear family system[10] over the past four to five decades, along with population growth has contributed to the increase in the number of houses being built on such land. While the population of the Kollengode panchayat has increased by 26.2% during the three-decade period between 1961 and 1991, the number of occupied residential houses has increased by 32.2%. In Elavenchery, the increase has been sharper. While population increased by 59.5%, the number of occupied residential houses has increased by 73.4% during the 1961–1991 period (Census Reports). While farmers report that the increasing density of houses has removed the existing tree cover on elevated lands, there exists no data to substantiate such observations.

The increasing spread of cultivation of crops such as coconut, banana, pepper, and areca nut (classified as 'mixed crops' in the 2001 survey by the Kerala State Land Use Board) on hitherto parambu land has also led to the removal of wild vegetation on these lands. The cultivation of the above-mentioned crops covers 933.22 hectares and 391.25 hectares in Kollengode and Elavenchery panchayats respectively (33.71% and 19.78% of the total agricultural land in the respective panchayats) (Kerala State Land Use Board 2001). Of the 933.22 hectares of coconut and mixed crop cultivation in the Kollengode panchayat, only 1.7% (15.85 hectares) is on converted paddy lands, the remaining being cultivated on sloping parambu land. As indicated in Map 3.1, the area shown as mixed crop in the watershed is mostly concentrated in the elevated portions of the watershed, which is, in most parts, adjoining the main or the distributary canal. Along with house construction, the majority of mixed crop cultivation therefore takes place on parambu lands, indicating the extent of removal of the prior diverse tree cover from these lands.

Box 3.2 The Spread of Housing Clusters and the Loss of Forests

The Mookarshan Hill, with the southern and northern slopes draining into the Tachakora and Matacode valleys is a typical example of a forested hill getting converted into a bare housing cluster. The Kadukode pond is located at the base of this hill. Murukeshan recalls that when he migrated into this area thirty years ago, this hill was full of trees. The first round of felling took place when the erstwhile owners of the hill and the land around, the Kollengode Kovilakam, in anticipation of the nationalisation of forests sold the trees to wood traders. Later, as the land was given to landless agricultural labourers as house plots, the remaining trees were also felled. Large stone boulders, wild creepers and trees, growing on land wedged between house plots, are the only remnants of the earlier vegetation pattern. The boom in house construction, encouraged by the government sanctioning house loans has converted this hill into a housing colony.

Source: Field Notes, 19.4.2002

Map 3.1 Cropping Pattern in the Varayiri Watershed

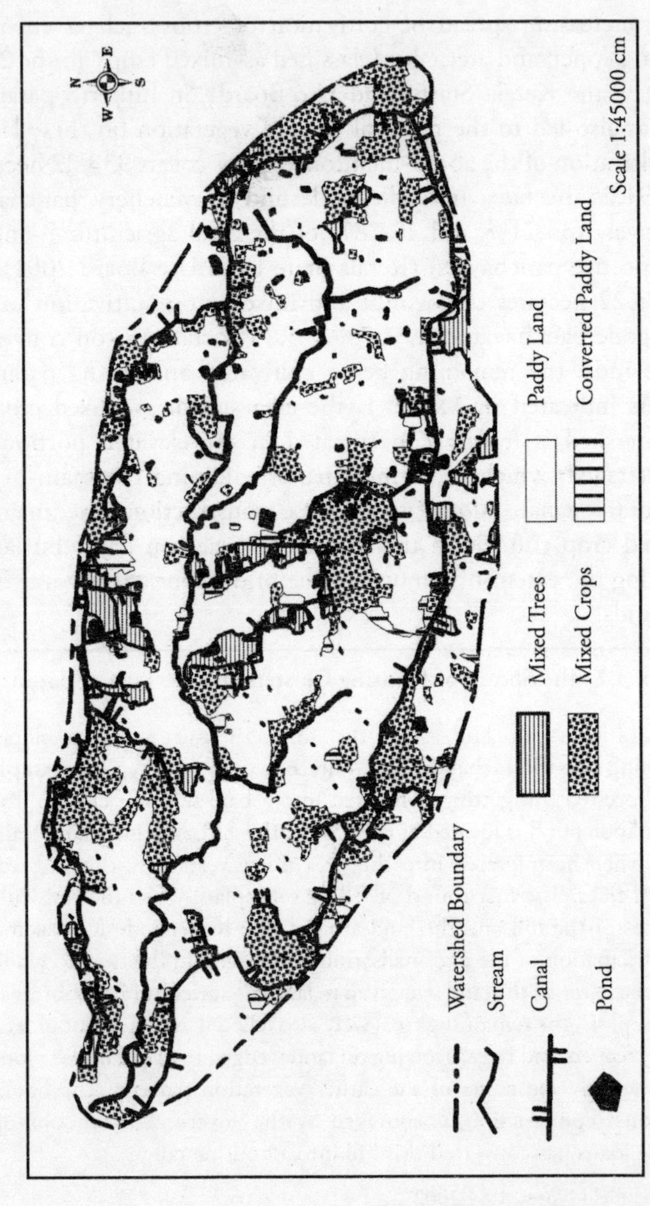

Source: Modified from Kerala State Land Use Board 2001

The above data indicates that the spread of mixed crop cultivation and the spurt in house construction, along with the nationalisation of private forests, has led to a decline in the natural vegetation cover of the area, particularly on the higher slopes of the dispersed hillocks in the area. The impact of the removal of tree cover on the infiltration of run off into subsurface zones, and its impact in turn on groundwater and surface water flows remains unstudied. When even large rivers like the Bharathapuzha, whose degradation has been highlighted (S. C. Nair 2004, Prabhakaran 2003), have not been subject to any detailed basin wide study that relates land use changes to river flows, it is no surprise that the degradation of small streams like the Varayiri goes unnoticed. Farmers in the upper reaches of the watershed almost never mention the plight of the Varayiri stream while discussing the problem of water scarcity in the area. To them, the stream flows far down below, and the diminishing flow in the stream does not affect them. Farmers in the lower reaches are directly affected by diminishing flows, but they do not correlate upstream changes in land use patterns with the present state of art of the stream. Many farmers attribute the drying up of the stream to a decline in the amount of total rainfall. This common place assumption that there exists a simple linear relationship between rainfall and stream flow has been the subject of detailed critique (Bandyopadhyay 1987: 2160), it being argued that it is the failure to maintain the ecological processes which allow rainfall to infiltrate and percolate to the underground to be discharged as perennial flows that causes the seasonal drying up of surface water sources (ibid.).

Either too little or too much

In contrast to the picture of the dry streambed for most parts of the year, is that of the stream flooding its channels during the monsoons. A week after the heavy rains set in, particularly during the southwest monsoon period (June–September), the Varayiri begins to flow. Due to the minimal need for irrigation at this time, farmers upstream keep their *kazhayis* (open mud valves) open, letting down the excess water in their fields. The absence of vegetation cover that can hold back the run off leads to flooding of the stream channel downstream. Once the monsoon withdraws, farmers close their kazhayis, thereby reducing run off into the lower reaches. According to farmers, a couple of weeks or at the most a month after the withdrawal of the monsoons, the flow in the stream diminishes. Alternating between little/no flow and flooding therefore is a characteristic feature of the streams in the watershed. As in the case of diminishing stream flows, farmers rarely correlate land use changes with the incidence of regular flooding. The silting of ponds also contributes to flooding downstream, as the ponds are unable to swallow the concentration of rainfall during the monsoons (Athreya et al. 1990). Flooding of the stream channel is a serious problem as far as farmers in the lower reaches are concerned, particularly in Valluvakundu, Poratancode, Tumbikode and Velampotta. Over flooding of the stream channel is particularly damaging when the paddy crop is in the flowering stage. Intermittent flooding also makes it difficult for farmers to apply fertilisers as the fertiliser gets washed away.

Stream as the carrier of canal water

As in the case of the pond, the degradation of the stream is most manifest in the fact that it primarily exists as a conveyor of canal water. Apart from the spate flows during the monsoons, the only time when there is any significant flow in the stream flows is when water is released through the canal network. Approximately two weeks after the dam has been opened in November/December, by which time water has been supplied to the upper reaches of the watershed, it gradually seeps into the streams below. This downward flow is facilitated by the fact that the canals (both the main canal and the distributaries) run through relatively high ground. During the initial two weeks, farmers upstream stock as much water as possible in their ponds and fields. When they can stock no more, the water is let down. Missing and leaking shutters also add to the leakage of water downstream. Canal water in the otherwise drying stream is a critical source of relief for paddy farmers in the lower reaches of the watershed. So much so that there are farmers in the lower reaches who ensure that shutters are kept open, so that water leaks into the stream. During the second crop, the fact that it is the water supplied through the canal system that is circulated through the stream and pond makes clear the extent of dependence on external water supplies.

The spread of tubewells

Investment in tubewells is yet another irrigation option for farmers who can afford the initial installation costs. The ability of groundwater irrigation in combining regularity of availability with convenience and easy accessibility (Raju et al. 2004: 263) has led to its growing spread[11] in many parts of the country. In the case of irrigation, they are becoming popular 'exit solutions' (Wood 1999: 775) largely due to the inability of the canal system in delivering water on a timely, adequate and regular basis (Molle et al 2003). Though one or two tubewells made their appearance during the 1970s, they have made their mark as sources of irrigation only since 1990. Until 1990, as per Census surveys, only 2 hectares of land in the Kollengode panchayat, and none at all in the Elavenchery panchayat was irrigated by tubewells (Census of India 1991). By 2005, there existed 119 and 149 tubewells in both these panchayats respectively (GOK 2006c). This is in addition to the tubewells that have been dug for drinking water purposes (a total of 105 tubewells in both the panchayats) (ibid.). The fact that tubewells are being dug for drinking water as well indicates the inadequacy of open wells in meeting the drinking water requirements of the population.

The average depth of the tubewells is 250 ft, though there are many which have yielded water at less than 200 ft and some which have not despite drilling to a depth of 400 ft. Of the 26 tubewells in the watershed, 19 are functional. Tubewells on their own are unable to meet the water requirements of paddy. However they are critical sources of supplemental irrigation. They are also vital to the irrigation of mixed crops, particularly coconut plantations, which are mostly located on elevated land where no other source of irrigation is available. While the canal network passes through the elevated portions of the watershed, water supplied through the canals is used only for paddy cultivation. In two cases in the watershed, water from the tubewells is pumped into ponds, which is then used for irrigating paddy.

The other sources that tap into the groundwater regime are the kuzhis (shallow pits discussed in Chapter 2). No new kuzhis have been dug in the area (most of them have been here for as long as farmers can remember). The only difference is that manual lifting has been replaced with energised lifting. In addition, some of the shallow wells, which are usually dug for drinking water purposes, are used for irrigating vegetables as well.

The 'Silent (Pump) Revolution'[12]

An important, yet un-investigated factor that affects the water storage and flows in ponds and streams in the watershed is the rampant lifting of water for irrigation. The increased use of energised lifting devices is indicated in the district level data on the net irrigated area (source-wise). During the period between 1995 and 2005, while the area irrigated by government canals and ponds have declined, the area irrigated by private canals, private wells and tubewells have increased (GOK 2007a; GOK 2001b). Corresponding figures for the panchayats studied are not available.

The 'dissemination of relatively cheap pumping technology' (Molle et al. 2003: 1)[13], as in other parts of the world, has radically transformed water use patterns from surface and groundwater sources. Across the country, assistance provided by respective governments in the form of subsidised electricity pricing, subsidised supply of diesel/kerosene and liberal financial assistance for the purchase of the pumping devices have contributed to their wide dissemination. (Bhatia 1992, Dubash 2007, Reddy 2002). During the 1970s, in a bid to encourage pumping of water from small water sources, the government of Kerala not only devised a scheme for the free supply of pump sets but also recommended the urgent energisation of the same, even calling for special budgetary provisions by the State Electricity Board if necessary (SPB 1976:4).

The dissemination of low-lift pumps have enabled the easy pumping of water from streams and ponds. Most of the pump sets used have a pumping capacity of 1–5 HP. In the case of tubewells, pump sets of 5–10 HP are used. Pump sets have facilitated individual access to water, providing much needed flexibility and individual control in the application of water as compared to the bureaucratically managed water distribution practices of the Chulliar canal system. Individually operated pump sets are more in number here,[14] as compared to lift irrigation schemes, which irrigate fields belonging to a group of farmers.

During the second crop season, pumps are found to work non stop in the lower reaches of the watershed where canal water supply is inadequate (being located in the tail ends), lifting out water from the Varayiri stream and the Gayatri River. Prabhakaran, a farmer from the Valluvakundu area remarked that during the second crop season, one would see a row of diesel engines along the stream, pumping the stream till it was dry.[15] The increasing intensity of water withdrawals from streams and rivers is often justified by the misplaced logic endorsed by many a riparian landowner that water that flows down is 'wasted', and hence should be used as much as possible. As one such farmer from the Peringotukavu area remarked,

> In all streams and rivers, water is running wastefully into the sea, and it is of no use to anybody.[16]

Neither the panchayat (in whom control over streams is vested) nor the minor irrigation department, in whom control was vested prior to the decentralisation of resources to panchayats in 1996, have initiated any action to assess the cumulative impact of individual withdrawals from the stream.

Apart from individual withdrawals, the implementation of lift irrigation schemes irrigating 50–100 hectares is exerting added pressure on declining flows in streams and rivers. Amongst the lift irrigation schemes, the Tootipadam lift irrigation scheme has been functional since the early 1990s. Two lift irrigation schemes across the Gayatri River were being proposed when the study was conducted, viz. the Peringotukavu and the Tumbidi-Karippayi lift irrigation schemes. The intended command area of both these schemes was 64 and 52 hectares respectively. Like the Tootipadam scheme, the latter two also aimed at irrigating those areas that did not receive canal water supply. However, unlike the Tootipadam scheme which lifted water from the Gayatri River and supplied it through pipelines, the latter two planned to store the lifted water in some of the existing ponds and then distributing it through the laid out field channels of the Gayatri irrigation project.

In the case of ponds, the pumping of water from ponds to potta and kalayi lands alike is reported to have reduced water levels in many ponds. This has primarily come about through the initiation of double cropping of paddy as well as the cultivation of coconut trees on potta lands and adjoining parambu land. In the case of the latter, farmers holding water rights to a pond, lift water to irrigate the coconut trees on potta lands as well. This is particularly so during the months between January to April, when rainfall is minimal. This contributes to reduced storages in the ponds, which affects the water levels in nearby wells and kuzhis as well. The issue of reduced storage is partially masked by the periodic supply of canal water into ponds during the second crop season when the need for irrigation is at its peak. However, the increasing unreliability in the supply of water through the canals has brought home more sharply the fact that the storage in ponds is fast declining, particularly in ponds located in the tail end of the canal network.

Intensified pumping of water is manifest not only in the increasing number of tubewells and the number of pump sets, but also in the changing cropping pattern of the area. The growing spread of coconut, banana, pepper, and areca nut cultivation in the study area indicates the same. The cultivation of these crops requires intermittent irrigation, which intensifies during the relatively rainless months of January–April, when flows and storages are at their lowest. While the availability of energised pumping facilities promotes the cultivation of these crops, the cultivation of these crops also leads to an increased use of energised pumping, leading to a vicious cycle of changing cropping patterns and increased water consumption. In addition, lifting of water is being increasingly resorted to during the second crop paddy season, when the availability of water through the canal network is highly unreliable. The pumping intensifies during the relatively rainless months of January–March. As discussed in Chapter 5, the unreliable supply of water to the second reach of the canal ayacut, causes the second crop of paddy to extend up to March.

It is difficult to arrive at any estimate of the total volume of abstractions from ponds, streams and wells. Neither is there any quantitative understanding on how pumping from one source affects water flows in other sources. With regard to groundwater exploitation, it is quite likely that many a shallow well and bore well tap into the same shallow aquifer, for farmers from certain areas report that the drilling of a few tubewells in close proximity has led to the lowering of water levels in shallow wells in the same area. Madhu from the Tachakora area for instance remarks that the water in the open well in his house compound never used to dry until a few tubewells were dug in the vicinity. Similarly, Mukundan in the low lying Velampotta area observed that the digging of two tubewells by the neighbouring landowner for irrigating his coconut plantation, led to falling water levels in his pond and well, so much so that he had to dig another well for drinking water purposes. The continuing digging of tubewells is however unmindful of the hydrological inter connections between shallow and deep wells. While there are no clear estimates of the extent of groundwater withdrawal from the panchayats studied, it is pertinent to note that the adjoining Chittur Block Panchayat has been notified as overexploited by a recent estimate of groundwater withdrawals in the state (Bijoy 2006).

Groundwater exploitation could also have an impact on stream flows and vice versa too, when streams are hydraulically connected to groundwater aquifers (Glennon 2002, Burke et al. 1999). In such cases, water will move between stream and aquifer depending on the water table in the surrounding aquifer, with streams contributing to groundwater recharge when levels in the aquifer are lower than the stream, and receiving water when water levels in the aquifer are higher.[17] Pumping from either of these sources therefore affects water levels in the other. In the Varayiri context, it needs to be established how many streams are hydraulically connected with local aquifers. Such an insight would help to assess the likely impact of intensifying groundwater extraction on declining base flows in streams and rivers, particularly during the summer months.

Regulating withdrawal

The widespread use of pumps therefore poses an important challenge to the regulation and management of water resources (Molle et al 2003: 1). Individual pumping of water from public water sources such as streams and rivers for irrigation purposes is forbidden as per the existing irrigation laws. This restriction can be enforced by prohibiting the use of pump sets by individuals for irrigation purposes. While the operation of electric pump sets can be regulated by withholding the departmental sanction required for the same,[18] this is not possible with regard to the use of diesel pump sets whose installation requires no official permit. In the case of electric pump sets, the restriction on individual pumping is evaded by manipulating the public-private categorisation. In order to acquire the necessary permit for the installation of an electric pump set from the agricultural officer, farmers dig wells on riparian land and depict the privately owned well as the source of irrigation and not the stream. Theoretically, water in the well is private property as it is located on private land, but in reality, it is the water in the stream that seeps into the well. Once the agricultural officer testifies that the concerned water source is privately owned, the Kerala State Electricity Board gives the sanction for the installation of the electric pump set. Bribing of the concerned officers for speedy processing is often resorted to. This leads to a situation wherein well to do farmers are able to install electric pump sets on riparian land, freely pumping water from streams and rivers. In the case of tubewell installation, a sanction is required from the Revenue Divisional Officer (RDO). While obtaining a sanction for pump sets below 3 HP is easy, sanction for pump sets with a higher horse power is not so. That apart, the digging of tubewells per se is not regulated, it being left to the discretion of the individual landowner.

With regard to surface water planning in Zimbabwe, Vincent discusses how river flow records, along with calculations of rainfall, evaporation and runoff are useful in giving an idea of the limits on allocation of water permits given around the year (Vincent 2003: 111). In the case of lift irrigation schemes where electric pump sets are used to pump water from streams and rivers, the sanctioning of such schemes by the panchayats is not based on an assessment of flow records, implying that schemes can be sanctioned even when the flow is diminishing. There has been a marked increase in the number of such lift irrigation schemes across rivers and streams (see the following section). River flow measurements in the state are conducted only for large rivers and not for smaller tributaries and streams like the Varayiri. This is also because hydrological investigations in the country have been mostly viewed as a tool for assessing resources that can be made available by the implementation of surface water based major or medium irrigation projects. That it can be used as a tool to assess the nature and quantity of total water resources in a small watershed or a region or a river basin, and that it can also be used to assess the real physical impact of each development project or human intervention on the availability and distribution of resources has been largely ignored (Santhakumar and Rajagopalan 1993). The neglect of small water sources like streams and ponds from any kind of flow measurements makes it difficult to assess the kind of pressures that might be building up on these sources, which has important implications as far as local water scarcity is concerned. Vincent observes a similar trend in the Zimbabwean context as well, wherein hydrological data collection was confined to perennial rivers, which were considered to be 'public' water sources. Run off in lakes, ephemeral streams, and in intermittently flooded depressions on private land were not subject to any kind of formal assessment as they were regarded as 'private' water sources, the use of which was deemed to rest with the owner of the land. This legal demarcation of water sources into public and private, led large areas of land and water to remain outside the limits of hydrological analysis (Vincent 2003).

Assessing locally available water supplies and the total demand for water in each region holds relevance beyond the Varayiri watershed. In all parts of Palakkad district that are irrigated by water supplied from distant reservoirs, the deteriorating functioning of reservoirs places in doubt the long term availability of 'external' water. Locally available water resources therefore are assuming critical importance in meeting the water requirements of each region. It is therefore timely to assess the total water availability in each region and to ensure that water consumption does not threaten the long term sustainability of the water sources concerned.

Local Level Water Resource Planning

The above sections have illustrated the disregard for ecological integrity in the conventional approach towards the management of water resources. During the last decade, local water resource development and management has been emphasised as an important strategy in meeting future water needs, particularly in water-scarce regions in the state (GOK 2006d). This has received added emphasis with the implementation of decentralisation policies in the state since the mid 1990s, whereby local institutions of self government (the panchayats) have been vested with additional decision making powers in local water resources planning and management. This has also coincided with a gradual shift in the state's irrigation policy with a greater emphasis being placed on minor irrigation development (see chapter 1), which is focused on water sources such as streams, ponds and wells. In the following sections I look into whether the contemporary emphasis on local level water planning and management has brought about changes to the existing infrastructure and supply-oriented approach towards the management of water resources.

As per the Kerala Panchayati Raj Act of 1994 (the state level legal enactment following the 73rd and 74th Constitutional Amendments that devolved enhanced powers and responsibilities to local level bodies in the country), all rights and liabilities of the government in relation to public water sources are to be vested in the village panchayat.[19] Public water courses have been defined to include the beds and banks of river streams, irrigation and drainage channels, canals, lakes, ponds, cisterns, fountains, wells, *chaals*, stand pipes and other water works, including non-private land which lies adjacent to these sources. It excludes rivers passing through more areas than the panchayat area (Kerala Panchayati Raj Act 1994). All minor irrigation works (defined as irrigation works having a cultivable command area up to 2000 ha) were vested with the panchayati raj institutions (GOK 2005). The Grama Sabha (the village electorate) has been empowered to suggest the location of community water taps, public wells, public sanitation units, irrigation facilities and so on, as well as to identify deficiencies in water supply and suggest remedial action (Kerala Panchayati Raj Act 1994:40). The mandatory functions of the village panchayat (as listed in the Third Schedule of the above mentioned Act), also include maintenance of traditional drinking water sources, as well as the preservation of ponds and ponds within its jurisdiction.

Higher levels of the panchayats were also mandated to intervene in watershed management programmes. While the Block Panchayat is entrusted with the management of watersheds falling within its jurisdiction, integrated watershed management programmes covering more than one Block Panchayat is the responsibility of the District or Zilla Panchayat. Similarly, the Block Panchayat is entrusted with the implementation and maintenance of all lift irrigation schemes and minor irrigation schemes that cover more than one village panchayat.

Hence, while major irrigation continued to be vested with the Irrigation Department, all minor irrigation works were transferred to the panchayats, including issues related to their management. This provision was however repealed by the enactment of the Kerala Irrigation and Water Management Act of 2003, which revised the definition of minor irrigation. As per this new Act, minor irrigation works refer to only those works that irrigate or are useful for the drainage or protection of a cultivable area of not more than fifteen hectares. All other works having a cultivable command area greater than 15 ha have been taken over by the Water Resources Department as medium irrigation. A considerable degree of confusion exists regarding the powers and responsibilities of the panchayat with regard to minor irrigation. While this has implications over local water resource development, the Act of 2003 had not come into force when the study was being conducted. Hence during the study period, the panchayat was empowered to initiate programmes and projects relating to local water resource development. In addition, the plans for integrated watershed management could be planned and implemented at the Block panchayat level as well. This was a significant development, as it implied that the three-tier panchayat system, if it wished to, could formulate locally appropriate water management plans. This was a challenge particularly in panchayats facing water scarcity.

Panchayats and minor irrigation

A perusal of the Vikasana Rekhas, the Development Plans generated by each panchayat as a part of the Campaign for Decentralisation during the Ninth Plan period (1997–2002) reveals the underlying approach towards water resources management. These plans gave an outline of the state of art situation with regard to agriculture, irrigation, drinking water, health and housing in each panchayat. With regard to water and irrigation, each panchayat has listed out the water sources within its jurisdiction, the problems that they face and the proposed plan of action for the future. The development plans for the Kollengode and Elavenchery panchayats revealed that the action plan for irrigation and water resources management was focused on the identification of projects that would increase the amount of water for irrigation and drinking water purposes. The Vikasana Rekha (the Development Plan) of the Kollengode Panchayat for instance, put forth twenty-one recommendations in the water resources sector. Ten of the recommendations proposed construction of check dams across various streams, of which a sizeable number (seven) proposed a lift irrigation scheme along with the check dam. Check dams are low dams, measuring 3–5 metres in height, constructed across the streambed. They are constructed with the intention of slowing down the flow in rivers and streams, so that the water that accumulates behind the dam replenishes the water levels in nearby wells, and thereby abating drinking water scarcity during the summer months. While check dams are portrayed as water conservation measures, they enable the extraction of water for irrigation purposes as well. An additional three proposals focused on renovation of existing check dams. Three of the proposals recommended supply augmentation of the storage in the Chulliar reservoir through the speedy implementation of the earlier discussed Palakapandi and Kuriarkutty-Karapara diversion projects as well as the deepening of the Meenkara-Chulliar link canal. There were two proposals for the construction of a 'mini dam' below the waterfalls in the hills for irrigation purposes. Others included construction of sidewalls along the stream, deepening of ponds, renovating the canal network and so on.

Keeping aside the proposals for supply augmentation, most of the remaining proposals were therefore focused on construction of check dams and associated lift irrigation schemes. The heavy focus on the latter was primarily due to the fact that it was minor irrigation that was vested with the panchayat. The implementation of these schemes was made possible through the enhanced flow of funds to local bodies as a result of decentralisation, as well as through the special support received from external agencies like the European Economic Community, the Dutch Government, assistance from NABARD and so on (GOK 2005). During the Ninth Five Year plan, a total of forty check dams had been envisaged across the Bharathapuzha and its tributaries, in addition to the already existing thirty two ones (DPC n.d.). This emphasis on lift irrigation and other minor irrigation schemes by the panchayats has been reported from all over the state since the implementation of decentralisation. A review of panchayat spending on irrigation during the first two years of the Ninth Plan (1997–98 and 1998–99) reveals that while the state government allocated rupees two hundred million a year towards minor irrigation, local governments across the state set aside almost four times the same amount (i.e. rupees seven hundred and sixty million) (GOK 2002a).

While there has been an explicit focus on implementing minor irrigation schemes, at no point in the Vikasana Rekha has the issue of sustaining water flows in streams and rivers been mentioned. Check dams, by virtue of blocking surface flows and thereby leading to a slight enhancement of the water table in the surrounding areas, have been considered as symbols of water conservation. In most cases however, the water that collects behind the check dams is mostly used for water consumptive irrigation.[20] The more important issue of forest conservation in the catchments of streams and rivers, with a view to enhancing water flows is not considered (S. C. Nair 2004).

In addition, the cumulative impact of check dams and lift irrigation schemes on downstream flows has not merited any attention. It needs to be noted that none of the streams across which these structures have been proposed are perennial. Whenever downstream users have protested against the construction of a check dam or the implementation of a lift scheme upstream, they have been pacified with the promise of a check dam in their area. During the study, a check dam was being proposed at Puzhapara across the Tekkepuzha River. The check dam was proposed to be constructed at a total cost of Rs 4 lakhs by the Palakkad District Panchayat. The people were planning to seek funds from the *Nelkrishi Vikasana* Agency (Paddy Development Board) for the construction of the lift irrigation apparatus at the same site. When downstream protestors argued that their pump sets would go defunct, they were pacified by the panchayat by assuring them that another check dam would be constructed for them downstream. This trend continues despite scarcity of water being identified as one of the major problems faced by lift irrigation schemes (SPB 1998). Such an approach can lead to what Molle defines as an 'over commitment' of resources, that is fostered by a flimsy knowledge of hydrology (Molle 2003).

Panchayats and water conservation

This section explores the extent to which water conservation programmes taken up by the panchayat meet the intended objective. Water conservation has been declared to be an area of priority by both the panchayats of Kollengode and Elavenchery. Apart from the construction of check dams, pond restoration has been one of the activities taken up in the name of water conservation. This has been the case with all the panchayats in Palakkad district, which exhibit a high density of ponds. Pond restoration has however been equated with the construction of stonewalls along the insides of ponds. The justification for the same is that pond bunds have been prone to erosion during heavy rains, particularly so as regular maintenance of the bunds is no longer undertaken. The 'side protection walls' as they are commonly referred to, are considered to protect the sides of

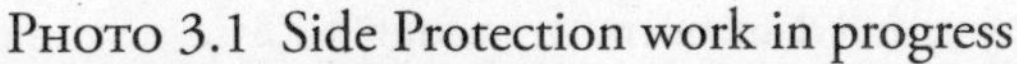

PHOTO 3.1 Side Protection work in progress

the pond from caving in. Once the stones are packed together, the outermost upper layer is lined with a layer of cement.

Such attempts at pond restoration conform to the 'classical' approach to pond rehabilitation that focus on structural and high cost measures (Shah and Raju 2001). Shah and Raju emphasise the importance of undertaking pond restoration or rehabilitation only after an ecological profile of the pond has been established (ibid.). The basic components of this ecological profile would be the state of art of the catchment, the land use patterns there in, the rate of soil erosion and gulley formations in the catchment and so on.

Establishing an ecological profile of a pond would lead to a situation wherein restoration of the pond catchment is given as much emphasis as the restoration of the pond per se. Very often erosion of the pond bund lies at the end of a long path of erosion points in the catchment of the pond. Merely constructing a stone bund without addressing the problem of soil erosion in the catchment is not a long lasting solution. Land use planning in pond catchments which is important in ensuring water flows (both surface and sub surface) into ponds, is critical. Many farmers observe that the increasing construction of houses in pond catchments has reduced the water flow into the pond. Similarly, the reduction in the tree cover in almost all pond catchments and its impact on sub surface flows into the pond has not been considered important enough to assess. The possibility of more cost effective measures such as vegetative protection of bunds is not considered as an alternative to such high cost measures, despite farmers recalling the wide varieties of trees that used to grow on pond bunds in the past.

Yet another activity that has been categorised as a water conservation measure is arresting the degradation of the stream channel. Flooding of the stream channel during the monsoons or during periods of heavy downpour has resulted in the stream bund caving in at many locations. A number of flood protection schemes have been implemented in the area, which seek to protect the stream bund through the construction of stone protection walls (similar to the type built along the insides of the pond) along the stream margins and thereby prevent flooding. Once again, rather than address the causes of flooding, efforts are focused on the terminal points of damage viz. the stream channel. This is primarily because the river or the stream channel is not viewed in conjunction with its catchment area. Arresting ecologically harmful land use practices in the cathcment area is for instance not viewed as a stream protection measure.

Exclusive emphasis on such structural measures carries other negative implications as well. The resultant decrease in the width of the stream channel, due to the construction of stonewalls inside the existing channel, has been reported by some farmers to increase the incidence of flooding. Stone banks have also been reported to increase the velocity of flow. The ineffectiveness of this method of flood control has been reported from other parts of the world as well (Glennon 2002, Purseglove 1988). Glennon observes that this form of flood control aggravates the situation as river channels are constricted by the lining of soil cement, which funnels floodwater downstream with greater velocity and force. He also notes that the greatest erosion and flood damage occurs at whatever point downstream that the soil cement ends. Soil cementing of riverbanks has also been shown to seal the fate of riparian vegetation, bird and animal life as well (ibid.).

This was illustrated when the sides of a small drainage channel carrying the drainage from paddy fields in the Maripadam area was cemented by the Kollengode panchayat under a flood protection scheme during the second half of the 1990s. While water is reported to have moved down very slowly through the earlier grass laden channel, it was found to drain away very fast through the cemented channel. Farmers report that apart from reducing soil moisture levels of adjoining fields, the cementing of the channel increased the incidence of flooding downstream, during the monsoons. Moreover, as in the case of the pond, the construction of stonewalls, has been found to destabilise the existing stream bank vegetation.[21] Apart from the method chosen to prevent flooding, the ad hoc nature of implementation aggravates the matter. Side protection works are not implemented along the entire stream channel, but in parts as the high cost involved makes it difficult to build walls along a long stretch. Constructing the wall in bits and parts creates problems in the flood prone areas, for overflow tends to be severe immediately downstream of the point at which walls have been constructed. Similarly, at times, the wall is constructed along one side of the channel only, leaving the other side more vulnerable to the erosive power of floodwaters. To make matters worse, most of the time the selection of the site is determined by political and not ecological considerations.[22] In some other cases, the work has been implemented in sites that do not require flood protection.[23]

In a nutshell, panchayats have expressed an inclination to take up such structural measures in the name of water conservation. Non-structural measures that seek to promote surface and sub surface recharge through enhancing the vegetative cover for instance, or even vegetative protection to pond bunds and stream banks to prevent erosion have not been given much emphasis. The panchayats however cannot be held solely responsible for such a narrow vision. As an outcome of decentralisation policies, panchayats were vested with decision-making powers in areas that have been traditionally managed by government departments. In the case of water resources management, they have only followed the approach adopted by the irrigation departments so far.

Keeping in line with the contemporary national and global policy emphasis on a river basin approach towards the management of water and other natural resources, government guidelines for the implementation of the ongoing Eleventh Plan (2007–2012) in the state emphasizes a watershed based approach to local planning (GOK 2007c). The Plan Guidelines propose that working groups be constituted at the panchayat level in order to facilitate the preparation of watershed based local plans. These integrated water management plans at the watershed level are to be prepared with 'a full understanding of the River Basin issues' (GOK 2007c),[24] and are to be merged to form a River Basin Management Plan. As per this recent directive, these watershed plans are to provide the framework for all development interventions undertaken by the local bodies. For instance, intervention plans in agriculture and allied sectors, as well as in irrigation and environment need to be based on watershed plans (ibid.). The stated objective of this new planning methodology to be adopted across the state is the 'progressive restoration of water–land–biomass balance and improved livelihood opportunities based on sustainable natural resource management' (ibid.).

The attempt is aimed at ensuring a more integrated approach towards the management of natural resources at the local level. Such an integrated approach is in direct contrast to the prevailing sectoral approach that is endorsed by the functioning of various government departments. The current approach for instance is one in where water management, forest management, soil conservation and agriculture are viewed as separate compartments. Prior to decentralisation, while the minor irrigation department looked into minor irrigation, the soil conservation department looked into watershed management programmes and soil conservation activities. These two activities were viewed as separate, despite the fact that the same water system (a stream for instance) was affected by both minor irrigation schemes and watershed programmes. Such a fragmented approach has continued with the implementation of decentralisation as well. While the village panchayats have been entrusted with minor irrigation, the block and district panchayats have been entrusted with the implementation of watershed management programmes. In addition, the passing of the Kerala Irrigation and Water Management Act of 2003, which took back minor irrigation from the panchayats and vested it once again with the minor irrigation department has added to the confusion.

The recent attempt at integration through the emphasis on panchayat-based watershed plans needs to be viewed in such a context. While this policy directive is still to be implemented, such a reorientation of perspectives calls for an overhauling of the existing ways of managing water and related natural resources. The policy guidelines in this regard discuss in detail the institutional changes required to facilitate such an approach. There is for instance much discussion on the constitution of a 'watershed sabha' (meaning the collective of people residing within a watershed), and watershed committees.[25] There is not much discussion however on how such a change in vision and approach is to come about.

Conclusion

The underlying approach towards water resources development in the study area exemplifies the 'resource capture dimension' (Turton et al. 2001). This is reflected in the construction of reservoirs and inter-basin transfers that lead water into the Chulliar reservoir. Even the management of smaller water sources such as the ponds, wells and the stream is dominated by this capture mode. In the process, the ecological consequences of such an approach are sidelined. In the case of the ponds and the streams, the reasons for their present degradation do not merit any attention. While the absence of hydrological assessments make it difficult to ascertain the factors that could have precipitated this degradation, land use changes in the catchments of streams and ponds, siltation of the stream channel and ponds, as well as unrestrained pumping have been outlined as some of the important factors. The neglect of the ecological dimension is reflected in the rehabilitation measures adopted as well, which are confined to structural measures focused on the terminal points of damage, viz. the erosion of the stream bank or pond bund.

Ecological factors are also ignored while envisaging of measures that are intended to ameliorate water shortages. Scarcity of water is viewed as a problem of deficient supplies alone. The role of ecological factors in declining supply is not considered. When long term ecological restoration programmes are not conceived as a part of the solution to water scarcity (S. C. Nair 2004), short term, easy to implement, supply augmentation schemes such as the proposed Palakkapandi project are the favoured solutions. These have political and popular appeal as well.

Despite the degradation of local irrigation sources, day to day irrigation needs are being met (though not always fully) through the supply of water from the Chulliar dam. This partly explains why the degradation of local water resources is not taken very seriously. Burke argues that the expansion of irrigated agriculture in the 20th century through engineering efficiency has 'decoupled the water user from the inherent risk of exploiting both surface and groundwater resources' (Burke 2000: 123,136), sheltering the user and the food production system from 'hydrological reality and the inherent limits of land and soil-water systems' (ibid.). In the study area, the advent of large scale storage structures, mechanised drilling operations, and energised pumping devices (both diesel and electric) has enabled the exploitation and use of larger amounts of water than in the past. The availability of water from across basin through the canal system, and from underground aquifers through the tubewell and the pumps, has removed the water user from the hydrological reality of the drying up of more proximate surface water sources, namely the ponds and the streams. In addition, the peculiar situation wherein canal water is circulated through the streams and ponds during the lean season, has led to further devaluation of these water sources. As Reddy argues, the dependence on external water supplies often leads to the neglect of local water sources (Reddy 1999). Farmers in the area meet their water demands without paying heed to the degradation of local water sources.

This situation is however fast changing with the increasing unreliability in the supply of water through the canal network. The growing rift between the states of Kerala and Tamil Nadu over the sharing of river waters[26] increases the vulnerability of the Chulliar reservoir. The unreliability in canal water supplies will enhance the pressure on the local water sources in meeting the demand for irrigation and drinking water. The increasing numbers of tubewells and check dams cum lift irrigation schemes in the study area are an indication of the same. The increasing numbers of lift irrigation schemes have reduced downstream flows, manifest in an increasing number of contestations between upstream and downstream users (see p. 246 in Chapter 6). Similarly, the increasing number of tubewells is a potential threat to the water levels in shallow wells and in streams. Despite the potential threat to long-term sustainability, such measures are gaining prominence, particularly since the implementation of decentralisation policies in the state that has increased the participation of the local governments in local level water resources management. Rather than leading to a critical appraisal of the local level water resource base, decentralisation policies coupled with the shifting irrigation strategy of the state have resulted in the enhanced extraction of water from local water sources.

Notes

1 The remaining three reservoirs located in the Bharathapuzha basin, but within the territories of the adjoining state of Tamil Nadu are the Aliyar, Upper Aliyar and the Tirumoorty dams, constructed across the tributaries of the Chitturpuzha, the Aliyar and the Palar, which originate in the Anamalai hills of Tamil Nadu.

2 The Meenkara Dam with a storage capacity of 11.3 million cu.m. has a larger catchment area of 90.65 km^2.

3 The fact that the government issued a tender in 2002 to mine this sand from the reservoir bed indicates the seriousness of the problem.

4 Though the agreement was signed between the two states on the 29th of May 1970, it was given retrospective effect from the 9th of November 1958, when work on the project had begun (Ravi et al. 2004: 35). Included in the agreement was a provision for review of the terms and conditions once in every thirty years (ibid., 23).

5 The Chitturpuzha scheme essentially consists of the Moolatara regulator from where the water released from the Aliyar reservoir is apportioned to different parts of the Chitturpuzha ayacut. Located at the border of the state of Tamil Nadu and Kerala, 40 km downstream of the Aliyar dam, this regulator has a storage capacity of only 19 mcft (million cubic feet). The other small storages, which were large ponds built in the 19th century by well to do landlords, later incorporated into the Chitturpuzha Irrigation System after the construction of the Aliyar dam upstream, also have limited storage capacity. They include the

Kambalathara Eri and the Venkalakayam Eri, reffered to as balancing reservoirs (GOK 1998) with a storage capacity of 3 million cu.m. each. The other balancing reservoir, the Kunnambidari Eri has a storage capacity of 0. 833 million cu.m. only. The other smaller systems, which are now a part of the irrigation system, consist of small *anicuts* and run of the river diversions.

6 As per the agreement, the annual allocation of 7.25 tmc ft (thousand million cubic feet) by Tamil Nadu was to exclude unutilisable floodwaters (Ravi et al. 2004: 42). However, in practice, the Kerala government as well as the farmers in Kerala allege that the release of floodwaters is made to be a part of the annual allocation, implying an unfair deal for Kerala.

7 In a few such cases, farmers say that they or their predecessors appealed to the irrigation engineers and contractors to position the outlet at a convenient location.

8 The expression is borrowed from Nair, who used it while describing the present state of the art of the Bharathapuzha River (S. C. Nair 2004).

9 In tropical regions in particular, it has been argued that the foliage of wooded slopes acts as a parasol to the ground, breaking the force of rainfall, helping the ground to absorb some of the moisture, which then exercises considerable influence in forming sub-soil springs which helps to maintain perennial flows in springs (Brohier 1937 in Goldsmith and Hildyard 1984). Forests are therefore considered as checkers of soil erosion, protection being largely due to under storey vegetation and litter, and the stabilising effect of the root network. (Bruijnzeel 1990).

10 The shrinking size of the family is reflected in the changing average family size in the panchayats of Kollengode and Elavenchery during the 1961–2001 period. In Kollengode panchayat, it declined from 5.32 to 4.9 during the above-mentioned period, and in Elavenchery panchayat it declined from 5.37 to 4.8 (Census of India 1961, 1971,1991,2001).

11 Its rapid spread in many parts of the country has been so significant that 70–80% of the total value of irrigated production is estimated to depend on groundwater irrigation (Deb Roy and Shah 2002 in Raju et al 2004: 263).

12 The usage 'silent revolution' is borrowed from the paper by Molle et al on the silent spread of the pumping technology and how it revolutionised access to underground and surface water sources. The authors argue that the explosion in the use of wells and pumps for irrigation, domestic and industrial purposes in the developing world is often hidden from view. This they argue is largely due to the tendency

to associate irrigated agriculture in the developing world with canals, dams, tanks and reservoirs (Molle et al 2003).

13 Molle et al (2003) have attempted to capture the dissemination of pumps and wells across the world over the past few decades, and the way in which they have revolutionised the withdrawal of water from surface and groundwater sources.

14 The decline in the cost of pumps has facilitated their private ownership. Those who do not own pump sets, hire them when in need.

15 '*Tirichum marichum pump annu*' is what he said, meaning that there were pumps all along the stream bank.

16 Such an argument emerges from the belief that water in streams and rivers is intended for human use alone. It also emerges from the ignorance of the impact of reduced flows in streams and rivers on aquatic life, on saline water ingress in the coastal areas and so on.

17 See Glennon 2002 for a description of 'gaining' and 'losing' streams in this context.

18 The procedure is as follows: the farmer who wises to install an electric pump set to lift water from a private water source such as a pond or well, renders an application to the office of the Agriculture Department, i.e. the Krishi Bhavan. The Agricultural Officer either visits the site or deputes another official to do so. The official concerned is expected to make a recommendation for the installation of the pump set after testifying that the water source in question is not a public water source, and assessing the extent of land to be irrigated. The recommendation that the farmer concerned be allowed to use a pump set of a particular horse power is made after assessing the water requirements of the land to be irrigated. This recommendation is then forwarded to the Electricity Department, which then gives the sanction for the purchase of the pump set. Farmers report that in getting the sanction for the pump set they have had to bribe officials in both these departments. When farmers have not bribed the officers in some cases the sanction for the pump set was delayed without any reason.

19 Section 218 of the Kerala Panchayat Raj Act (KPR) 1994.

20 'The intention is to create a series of water pools after the cessation of rainfall, in December, so that such water can be drawn off for water supply and in certain cases even for agricultural purposes, alongside the location of such check dams' (Bhattathiripad 2003).

21 In the process of constructing the sidewall along the stream at Valluvakundu for instance, a deep-rooted tree was removed from the existing bund with considerable effort.

22 At Velampotta for instance, a sidewall was constructed only along one side of the stream channel. This side fell in the jurisdiction of the twelfth ward of the Kollengode panchayat, and the panchayat member (the elected representative) owned the adjoining land. Hence, he was able to push through the case of constructing the wall adjoining his land. Downstream of the cemented side, there is a two-metre drop in the streambed, and Rauttar who owns land at that spot was anticipating increased erosion the next monsoon, due the construction of the sidewall immediately above his land.

23 When a proposal for streamside protection was submitted to the Elavenchery panchayat in 2001, an error was made in the writing of the panchayat ward number in which the work was to be implemented. Though the error was noticed before the implementation commenced, the panchayat refused to make the necessary correction, saying that the scheme had already been sanctioned for that particular ward, and hence could not be reverted. This was despite the fact that the problem of erosion was more severe in the ward in which it was originally intended.

24 Refer Government Order, G.O. (MS) No. 128/2007/LSGD dt. 14–05–2007 (GOK 2007c).

25 To prevent institutional overlaps, members of existing institutions and organizations in the watershed area, such as the panchayat members (at the village, block and district level), presidents/secretaries of the padashekhara samitis, of various agricultural organisations, presidents of primary co operative and milk cooperative societies, and so on are to become members of the Watershed Committee.

26 This is manifest not only in the lack of consensus between the two states on the amount of water that is due to Kerala, but also in the ongoing struggle of both the states over the century old Mullaperiyar dam across yet another inter-state river, viz. the Mullaperiyar River in south Kerala.

4

Property Regimes and Rights to the Use of Land and Water

The focus of this chapter is on the changing rights regime to land and water in the study area, and its implications on equitable access. The chapter discusses how the reorganisation of land rights through the implementation of land reforms has shaped access to land and water in the region. In order to do so, it has drawn on both secondary and field information of the complex changes in rights to land and water. While the first part of the chapter looks at changing agrarian relations that have shaped rights to land, the latter half looks at how rights to land have shaped access to water based on the field information collected. Apart from the equity implications, this chapter has also illustrated the sustainability implications of the splitting up of land and water into private parcels. It therefore argues that the reorganisation of land rights paid little attention to the requirements of long term sustainability of the agricultural resource base.

Property Regimes in Land and Water

Given the long historical interrelationship between land and water rights, it has been argued that efforts at addressing agriculture related water needs must take into account the complex inter-linkages between land tenure and water rights (Cotula 2006). In most customary traditions on land and water use, the right to use water has been dependent on the use or ownership of land (Hodgson 2004). The riparian law of water rights operational in many countries is for instance, dependent on the ownership or use of land adjoining rivers and streams. Similarly, rights to the use of groundwater are linked to the use or ownership of land under which it is located.

Land therefore has been considered as the most important property variable that determines the property character of irrigation water (Abeyratne 1990: 20), it being suggested that while examining property in water, and in particular property in irrigation, a 'back door' approach through the examination of land rights be adopted (ibid.). While in most situations, particularly in South Asia, access to land extends access to water, the reverse has also been illustrated. Studies indicate how controlling access to water points in the grazing pastures of Niger in Africa, gives de facto access and control over the surrounding grazing lands as well (Peters 1984). Customary norms in this region restricted access to water points in order to restrict livestock access to grazing lands (ibid.).

Land rights are defined as the system of rules, rights, institutions and processes under which land is held, managed, used and transacted (Cotula 2006). They specify who is to get access to land at which time and in which place (Lane and Moorhead 1995 in Dijk Han van 1996). Land rights include ownership as well as a range of other land holding and use rights such as leasehold, usufruct and servitudes (Cotula 2006). Water rights are mechanisms through which users access water for a particular use without jeopardizing another users' right to the same (Van Koppen et al. 2004). Water rights have also been viewed as a type of property right that in conjunction with land rights assigns access, use, liability and control over water to some person and social groups relative to others (Wescoat 2002). Given the very many varied forms and functions of water, it has been suggested that water rights can never be more than an 'umbrella concept', which includes a variety of different rights to different kinds of water (Benda Beckmann et al. 1996). In some places water rights are tied to the land, whereas in others they are fully severable (Cotula 2006).

Changing agrarian relations in Kerala

This section describes changing land relations in the state and in the study area that help to understand the present configuration of land

and water rights. The changes brought about by the implementation of land reforms in the distribution of rights to land and water become clear when compared with the pre reform situation.

During the period of colonial rule, the Kollengode and Elavenchery panchayats were a part of the Malabar district of the erstwhile Madras Province. Land relations in this region, as in other parts of the state were dominated by land lease rather than the ownership of land.[1] As in other parts of the state, individuals with land owning rights were known as '*janmis*' (referred to as landlords in most academic writings on this issue). The word '*janmam*' means 'birth', and in the context of land relations, janmam rights imply hereditary rights or birthrights to a piece of land (Ganesh 1991: 300). The janmi was connected to the actual tiller of the soil through various kinds of lease/mortgage arrangements, mediated by a number of intermediaries. Different classes of leaseholders held lands vested with the janmi on payment of a share of the produce (referred to as *patam* or *varam* in Malayalam). '*Kudiyar*' being the generic term applied to different types of leaseholders (Ganesh 1991:302), the term '*janmi-kudiyan*' was commonly used to refer to the relationship that governed land relations in the state. This system exhibited regional variations, with the form and conditions of tenancy differing in the three erstwhile districts of Travancore, Cochin and Malabar (which were merged to form the present state of Kerala in 1957).

Three broad categories of people were involved in land tenure arrangements, the janmis, the *kanakar* and the *verumpattakar*. While ownership of the land was vested with the janmis, they did not engage in cultivation directly, not even in supervision of the field and crops. They normally entrusted this activity to the *kanakkar*, who were the principal tenants, most of whom were again non-cultivators (Sardamoni 1982), on payment of a lump sum in money or in kind, and an additional nominal payment or rent along with other customary fees and dues. *Kanam* rights were usually hereditary (Frankel 1972). The Kanakkar were therefore mostly intermediaries, exercising control over the land owned by the janmis, by sub letting the land to various types of cultivating sub tenants, collectively known as verumpattakar. Hence while the janmi held customary rights over land, the kanakkar held the controlling rights, and the actual cultivation was undertaken by the verumpattakar. The kanakkar were therefore considered as superior tenants, and verumpattakars as 'bare tenants' (Radhakrishnan 1981) or tenants-at-will (Herring 1983). The hierarchy of land control and privilege roughly paralleled the caste-based hierarchy of social status (Herring 1990).[2] Most of the janmis comprised of princely families and Namboodiri and Brahmin families (see p. 38 in Chapter 1). Temple authorities also enjoyed janmam rights. In the erstwhile district of Malabar, two kinds of tenancy are considered to have been in existence. The kanam tenancy mostly in the hands of the Nairs, who were below the Namboodiris in ritual and social hierarchy, and who were relatively numerous too. The other tenancy was the verumpattam tenancy, which was mostly with the Tiyyas (or Ezhavas), another major social group, who were below the Nairs in the above mentioned hierarchy. Below them were the large group of landless agricultural labourers, on whom the tenants depended for agricultural work. They constituted the lowest rung in the caste hierarchy. In the study area, they largely came from the Cherumakkal caste group.

Despite this hierarchy between the groups involved in agriculture, it has been argued that land did not belong to any individual or group as private property (Radhakrishnan 1981). The different social groups involved are considered to have enjoyed different rights and interests in the land, which more or less corresponded to their position in the then existing ritual and social hierarchies. Ironically, the most oft quoted opinion in this regard is that of a British officer, William Logan, the Collector of Malabar in 1881, who had been appointed by the colonial rulers to study the prevailing land and agrarian relations in Malabar, particularly in the context of the prevailing agrarian discontent in Malabar.[3] Logan was of the view that 'joint proprietorship' in the soil existed in Malabar till then second half of the eighteenth century, wherein five hierarchical groups exercised rights to the land. This comprised of the janmi, various kinds of lease holders such as the *kanakkaran*, *verumpattakaran* and *cherujanmakaran* (comprising of artisan and other service caste groups such as carpenters, blacksmiths and washer folk), and the agricultural labourers. According to Logan, the first three groups divided the net produce from the soil after providing for the customary dues to the latter two groups. The rights and interests of all these groups were regulated and restricted by customary laws. The fact that no single individual or group exercised exclusive rights to the land is revealed by the fact that in the event of sale of land, the customary rights of persons holding rights as kanakkar, verumpattakar, as artisan and service castes, hutment dwellers and agricultural labourers had to be recognised (ibid.). This system of joint proprietorship is considered to have undergone changes in the second half of the eighteenth century, first with the introduction of direct revenue administration by the Muslim rulers of Mysore who conquered Malabar, and later further with the far reaching changes brought about by British rule towards the end of the eighteenth century (Logan 1887).

The introduction of the two broad categories of 'landlord' and 'tenant' by the British is considered as an attempt to simplify this complex and overlapping system of land rights (Vani 2002). Colonial misinterpretation of customary land tenure has been attributed to have caused an imbalance in power relations between the janmi and the various classes of leaseholders (Ganesh 1991, Prakash 1987, Sardamoni 1982). Logan for instance argued that the British misunderstood the position of the janmi as the absolute owner of the soil.[4] Such an interpretation is also argued to have been part of the colonial objective of maximising land revenues by creating a section of powerful, landed people who would act as their agents in the region (Prakash 1987). Hence, while the janmi came to occupy a position of considerable power in the decades preceding the implementation of land reforms, it has been argued that this was not always so.[5] The transformation of janmis from customary to statutory landowners enhanced their power and privileges, giving them the right to enhance rents payable by tenants and to legally evict them for non-payment of rent or on expiry of the lease period (Ganesh 1991: 321). Such a move is reported to have proved detrimental particularly to the small kanam and verumpattam holders of land (ibid.). Similar cases of misinterpretation of customary land tenure and land relations have been observed from other parts of the country as well (Neale 1962 in Hann 1998). The increasing power of the janmi prompted campaigns for tenancy reform during the colonial period itself. Legislations in favour of tenancy reform finally culminated in the passing of the land reform act in the post independence era in 1970, which did away with the janmi-kudiyan system of land relations.

Agrarian relations prior to land reforms in the study area

Amongst the three erstwhile districts (Malabar, Cochin and Travancore) of Kerala, it was in Malabar that the position of the janmi is considered to have been the most powerful during colonial rule. Menon for instance cites the ruling of the Madras High Court in 1916[6] in which the owner of the banks on either side of a river was vested with the ownership of non-tidal and non-navigable rivers as well, thereby implying that the cultivators were dependent on the landowners for the use of irrigation as well (ibid.). Malabar also had the highest percentage of tenancy in Kerala. According to the Survey on Land Reforms conducted by the government in 1966–67, owner cultivators in Palakkad district specifically, accounted for only 13% of the total agrarian households, while tenant households accounted for 79% (Frankel 1972). Similarly, the Malabar region accounted for the highest percentage (96%) of kudiyirippu tenancies as well. Kudiyirippu refers to the land on which the homes of landless agricultural labourers were built. They were given this land on lease by the landlords, and could be asked to vacate at the will of the landlord. In Palakkad district alone, 40% of the agrarian households comprised of kudiyirippu tenancies in 1966–67, indicating the high incidence of landless labourers in the district.

Ownership of land in the study area during the pre land reform era, was largely vested with Nair and Brahmin families. The tenants were mostly from the Ezhava family, though Muslim tenants were not uncommon. Certain temples also owned land in the area such as the Kachamkurishi Temple, in which case rent had to be paid to them. The Kollengode royal family, commonly referred to as the Kollengode Kovilakam was the most prominent land owning family in the region, who owned land right from Elavenchery in the west to Govindapuram in the east (which is today located close to the interstate boundary with Tamil Nadu). Apart from agricultural land, the forests along the hills to the south of these panchayats were exclusively owned by the Kollengode Kovilakam. Other well known land owning families in the area included the Panikkath family who owned considerable land in the Velampotta area, the Pazhayaveetil family who farmers recall owned land in stretches between Peringotukavu in Elavenchery panchayat to neighbouring Pallasana, owning 5–6 *kalaperas* (farm houses used for storing of threshed grain and other agricultural accessories) in the region, the Karimathil family who owned about 80 hectares of land in the Poricholam area, the Kozhisshery family and the Ravunniarath family. Many farmers also recall that the land they cultivated as tenants belonged to Brahmin families such as those of Venkitarama Iyer, Kailas Iyer and other Brahmin landlords who resided in various '*gramams*' (Brahmin villages) in the area such as the Alampallam Gramam and the Perumal Kovil Gramam.

Despite the possibility of evictions, like the kanakkar, the verumpattakar were also said to enjoy hereditary rights according to local custom. It was not uncommon therefore for customary relationships between landlords and tenants to extend over two to three generations. Many a hitherto tenant recalls farming the same piece of land for the past 50–100 years (it may be noted that land reforms were implemented more than thirty years ago, in 1971). Fifty-year-old Krishnadas from Manalipadam for instance recalls that his family has been farming the land at Tarapadam since the time of his great grand father i.e. for the past four generations.[7] All through the period, the land was under the ownership of the same landlord.

Tenants' rights to land included rights to both agricultural and forest land. The forests in the hills, which were almost exclusively owned by the Kollengode Kovilakam, provided an assured supply of fuel wood, green manure and fodder. People had to pay a small amount, known as *chungam* (meaning toll) to the guard appointed by the Kovilakam for this purpose.[8] A holding of about 4–6 hectares cultivated on lease would therefore normally consist of paddy land, some parambu land, a *kalam* (different from kulam which refers to a pond), for carrying out post harvest operations, and in many cases a pond, or more than one. This represented an average holding cultivated on lease. Given the close relationship between livestock rearing and agriculture (cattle required for ploughing and cattle manure required for fertilising the fields), each tenant would maintain a stock of cattle. The cattle shed was therefore an integral part of the landholding, a person being appointed to take care of the cattle in many cases.

The rent due to the landlord, referred to as 'patam' was paid at the end of each cropping season, with the account for each year being settled during the New Year festival of Vishu. Vishu not only marked the commencement of the cropping season, but it was also the time when contracts with tenants and agricultural labourers were renewed. In addition to the annually paid patam, the tenants would gift the landlord with other agricultural produce such as vegetables (including different varieties of gourds), bananas, tamarind and beaten rice, on special occasions such as Vishu and *Onam* (a festival that coincides with the harvest of the first crop in the months of August–September). In reciprocation, landlords would give the customary '*Vishu kaineetam*' (a token sum of money) along with other token gifts mostly consisting of agricultural produce (see Box 4.1). When tenants were unable to pay the agreed upon patam, there were cases when they would be asked to vacate the lands. Farmers also report cases wherein landlords were willing to accommodate a shortfall in the amount of patam due to them.

Despite tenancy being widespread, tenants did not comprise a homogenous category, there being both large and small tenants in this category (Radhakrishnan 1982). The Ezhava tenant family who cultivated the lands of the Parassery Kalam in Manalipadam owned by Harihara Iyer or the Kavungal family who cultivated lands in the Peringotukavu area which belonged to four Brahmins from Alampallam Gramam and the Natathu family in Chittepadam are some examples of large hitherto tenant households in the study area. In each padashekhara samity (registered bodies of paddy farmers cultivating an average area of 50–100 hectares), there are at least 2–3 such hitherto large tenant families, whose holdings have been subsequently subdivided. Members of the Kavungal family recall that the land they once cultivated on lease (which has been subsequently divided) covered half the area of what is known as the Peringotukvau *padashekharam* (cluster of paddy fields) today.

Hence, just as certain pockets of land were owned by particular landowning families, certain pockets were cultivated by large tenant families for a long time. Most of the large tenants also acquired ownership over such land and the kalam located on these lands during the time of land reforms. Kalams were a prized possession as they provided the facility for threshing, drying and storing of the grain. Following the land reforms, like the janmis in the past, many of the large tenants continue to be referred to as owner of such-an-such a kalam. Many of these large tenant families have had to surrender land in excess of the prescribed ceiling when the land reforms were implemented. On the other hand, there were also small tenants like Velappan, an agricultural labourer, whose father cultivated just one hectare of land, which belonged to the owners of the Velampotta Kalam.

During the early 1970s, agricultural labourers comprised the largest proportion (55% and 64% in Kollengode and Elavenchery respectively) of the total work force in both the panchayats (Census of India 1971). This was true for most parts of the district as well. Since most agricultural labourers were landless, this high percentage indicates the high proportion of landless agricultural households in the area. Agricultural labourers were involved in a wide range of agricultural operations ranging from ploughing, sowing, transplanting and weeding, as well as grazing the cows, collecting grass for the cattle and filling the ponds during the monsoons. Prior to the passing of various legislations that fixed the number of working hours for agricultural labourers during the 1960s and 70s, they used to work from the early hours of the morning to late in the evenings. Some of the older agricultural labourers recall being brought here from afar by the landlords of this area and were given land on which to build their houses (kudiyirippu).

Agricultural labourers and water management

The numerical majority of agricultural labourers in the study area as well as in the district on the whole led to an oversupply of labour, which was assumed to be one of the reasons why they maintained permanent working relationships with one landowner, reluctant to break the permanent tie, which was a source of job security and fringe benefits (Frankel 1972). This relationship was however mediated by the land in question. Labourers therefore continued working on the same land despite subsequent changes in ownership. They were also referred to as attached agricultural labourers (George 1987) or *sthira panikkar* (permanent workers) in local parlance. Following land reforms when the ownership of the land was vested with the tenants, the labourers continued to hold customary rights to work on the same piece of land. Such a practice continues into present times when sale of land implies that the new owner has to employ the same set of agricultural labourers. Custom forbade the farmer from employing labourers other than those who worked on his land on a permanent basis. The yearly contract with agricultural labourers was renewed annually on the occasion of Vishu.

Box 4.1 Renewing the Contract

Yesterday was Vishu. I observed the annual custom wherein farmers renew the annual contract with agricultural labourers who work for them on a permanent basis at Palakode. A lamp was lit, and grain was heaped on the floor. The farmer gave each labourer a share of paddy, coconut, oil, tamarind and *puli avara* (a variety of beans), along with 'kaineetam' (token money) folded in betel leaves. After the customary dues had been gifted, the farmer and the labourers went to the fields and crackers were burst. The field was ploughed and seeds were sown. This was considered an auspicious start to the season.

Source: Field Notes, 15.4.2002

The practice of employing labourers on a permanent basis has important implications on water management as well. The task of diverting rainwater (referred to as '*vellam tirikkal*') into ponds was an activity that heavily relied on the labour force of each farmer. With the first pre monsoon showers in April–May, labourers would be ready to go and divert the water into the ponds. They were even expected to go out in the night if need be, to direct the rainwater into the ponds. Many farmers opine that if labourers were not available for such work in the past, filling of the ponds would have been impossible. Labourers were often found to fight with one another over the diverting of water to adjacent ponds, which belonged to two different landlords or tenants. Similarly it has been reported that when the ponds were full during the peak of the monsoons, labourers would stand guard even at night, to let out water from the pond to prevent the pond bund from collapsing.

This system underwent changes with the land reforms and the labour strikes of the 1960s and 1970s. The militant labour strikes during this period (which has been attributed to the perception of the labourers that the main benefits of the land reform package went to the large tenants and not to the actual tiller of the soil) led to new agrarian legislations such as the Kerala Agricultural Workers Act of 1974. This legislation mandated a workday reduction from 12 or more hours to 8, tea and lunch breaks, a minimum wage, permanent reconciliation machinery, a provident fund for the labourers to which farmers had to contribute, and most significantly permanency of employment for attached labourers (Herring 1983). The growing strength of agricultural labour unions led to a situation wherein labourers were not on call as before. Labourers no longer stand guard over ponds, or ensure that ponds are full. In addition, labourers no longer associate their welfare with a good harvest as they are mostly paid in the form of wages and not in kind. Hence, they no longer need to go out of their way in ensuring a good harvest. It has therefore been suggested that the 'feudal' pattern of labour control that prevailed in the pre reform era, while being of an unjust nature was suited to hydraulic agriculture (Herring 1990).

Over the past couple of decades, the system of engaging permanent workers has also been gradually done away with. With the younger generation of agricultural labourers moving out of the agricultural sector and searching for other livelihood means,[9] there is a decline in the total labour force available. While agricultural labourers accounted for roughly 64% of the total work force in Elavenchery panchayat in 1971, the figure came down to 47% by 2001. The corresponding figures for the Kollengode panchayat were 55% and 42% respectively (Census of India 1971, 2001). During the same period, the proportion of workers engaged in non-agricultural occupations increased.[10] This means that each farmer has to manage with a reduced number of permanent workers, who are unable to finish the work on time. The permanent agricultural labourers associated with a particular farmer are also reported not to allow other labourers to come and work on the land to which they enjoy customary working rights. This often creates problems in terms of timely completion of weeding or harvesting. As a result, most farmers have terminated the services of permanent workers by paying them a lump some amount. The local office of the Marxist party often mediates such a process. Farmers have paid about Rs. 5000–7000 to each permanent labourer whose services they have had to terminate in such a fashion. Subsequently they engage the services of labourers as and when they require.

The implementation of the land reforms

Land reforms in the state were part of the larger wave in favour of redistribution of agricultural land in several parts of the world and the country in the post second world war era.[11] Due to the heavy concentration of land holdings in the state[12] and the low land-man ratio, Kerala was considered to represent one of the strongest cases for redistribution of land in India. The state also exhibited the highest percentage of landless households in the country (30.90%) in 1962.[13]

In Kerala, the case for land reforms had its precedents in the struggles waged for tenancy reform, which was headed by the relatively well to do tenants in different parts of the state (Radhakrishnan 1982). The struggle for tenancy reform demanded the abolition of rack-renting and forceful evictions by the landlords. A series of legal acts led to the promulgation of the Kerala Land Reforms Act of 1963 which was modified by subsequent amendments, finally coming into effect on 1–1–1970. Reducing landlessness and abolishing tenancy were the primary objectives. The three main components of the land reform programme in the state centred around conferring ownership rights on cultivating tenants for the land leased in by them, giving an option to homestead tenants (kudikidappukar) to purchase homestead land from their landowners, and finally taking possession of the surplus lands by the imposition of ceiling laws for distribution among the landless labourers and land poor farmers (Ramachandran V K 2000, Radhakrishnan 1982). As per the first provision, all rights, titles and interests of the landowners and intermediaries over holdings cultivated by tenants was vested in the government, and in so doing full ownership rights were granted to the cultivating tenants over land cultivated on lease. Tenants were expected to pay only a nominal sum as purchase price.[14] The purchase price was treated as a debt to the government and the tenants were not bound to forfeit their rights for a default in re-paying it (Radhakrishnan 1981). In the study area, it has been reported that many of the tenants had not paid up the full sum of money due from them.

As per the second provision, the act aimed at transferring ownership over homestead land to the *kudikidappukaran* (kudikidappukaran represented a special category of attached agricultural labourers), conferring on them the rights to purchase at concessional rates the small extent of land in and around their hutment.[15] Apart from granting ownership titles to the tenant and the kudikidappukaran, the third component of the Act aimed at bringing in some degree of equity in the distribution of agricultural land, through the fixing of an upper ceiling limit.[16]

Another closely associated legal intervention was the nationalisation of private forests in the state through the implementation of the Kerala Private Forests (Vesting and Assignment) Act of 1971. Surplus lands (land in excess of the ceiling) available for redistribution among the landless proved insufficient, as a result of which the government of Kerala decided to take over the private forests as a part of agrarian reforms, and to distribute a part of it amongst the landless (Chundamannil 1993: 64). As an outcome of this act, all private forests in the state hitherto vested with the landlords were taken over by the state, in order to assign it to agriculturists and agricultural labourers for cultivation, the objective being to utilise the 'viable private forests land' to 'increase the agricultural production and to promote the welfare of the agricultural production in the state' (SPB 1997: 12). Nationalisation of private forests in the state, as in the case of similar attempts initiated in many parts of the developing world was also an outcome of the belief in the ability of national governments to solve natural resource degradation problems (Bromley 1991:108, Bromley 2002). It was expected that scientific management of resources under government control would avert degradation (ibid.).

Land reforms in Kerala, despite being hailed as the 'most radical, comprehensive and far-reaching in South Asia' (Christodolou 1990: 145), has been critiqued by others for having failed to benefit the most needy, i.e. the landless.[17] An important factor that watered down the objectives of these two acts was the time lag experienced in their implementation, giving plenty of time to the landlords to evade the laid down stipulations. The Kerala Private Forests Act was in the offing since 1962 when a bill was passed for the take over of private forests. Since then most of the private forest owners were inclined to sell off the trees before it was taken over by the government (Chundamannil 1993), resulting in deforestation of a high order. Data on forest cover in Palakkad district reveals that forestland owned by private individuals reduced by more than half during the 1940–1970 period, i.e. from 970 sq. km in 1940 to 409 sq. km in 1970 (a 58% reduction). During the same period, forestland owned by the government also reduced by roughly 9%, reducing the total forest cover in the district by 35% (Amruth 2004). The forests in the study area which were mostly under the ownership of the Kollengode Kovilakam is reported to have been felled during the decade preceding nationalisation.[18] The trend of clearing forests for groundnut cultivation in the hills (on land leased out from the Kollengode Kovilakam) also intensified during this period. Trees were also felled to make charcoal, which was burnt as fuel in hotels and other small establishments. The removal of customary control over forest lands led to a period of uncertainty, when the forests were neither controlled by the erstwhile janmis, nor by the state government, validating Bromley's observation that while the nationalisation of forests by the state creates de jure state property, in practice it creates de facto open access (Bromley 1991).

Land Reforms and Equity

This section examines the equity implications of the implementation of land reforms. The first section examines equity with reference to the size of landholdings. The second section discusses equity from the angle of access to water.

Size of land holdings

The existing skewed land holding pattern in the area indicates that the implementation of land reforms has not made a significant impact on this front. As of 1995–96, approximately 80% of the individually operated holdings in the Kollengode Block Panchayat were below 0.5 ha (1.25 acres) in area, occupying only 10.41% of the total area (GOK 2001a). More important, 91% of the total holdings were below 2 ha in area, covering 33% of the total land area. In contrast, holdings above 2 ha in area comprised only 8.7% of the total number of holdings, but covering 66.25% of the total area.

TABLE 4.1 Numbers and Area of Individually Operated Holdings in Kollengode Block Panchayat in 1995–96

Size of holding in hectares	*Number of holdings*	*Percentage of the total number of holdings*	*Area in hectares*	*Percentage of total area*
Less than 0.5	17181	79.35%	1255	10.42%
0.5–1	1202	5.55%	870	7.22%
1–2	1379	6.37%	1940	16.10%
2–4	1276	5.89%	3545	29.43%
4–10	521	2.41%	2967	24.64%
Above 10	93	0.43%	1469	12.19%
Total	21652	100	12046	100

Source: GOK 2001a

Disparities exist not only in terms of the extent of land holding, but also in terms of the quality of land. Though large holdings in the area would not cover more than 6 hectares on an average (which is relatively small compared to large holdings in other parts of the country), the inequities between the large and the small/marginal farmer is considerable.

Part of the explanation for the present inequity in landholdings is related to the manner of implementation of land reforms in the area. Across the state it has been argued that while abolition of tenancy had been successfully carried out, identification of land above the ceiling limit and consequent redistribution was not effectively done (Ramachandran 2000). The evasion of ceilings by landlords and large tenants played a critical role in the watering down of the provisions of the act. Former tenants cite instances of landlords taking back the land leased out to tenants in anticipation of the reforms, partitioning the land amongst their children, resorting to bogus transfers of land to married daughters residing separately,[19] registering the land in smaller units within the land ceiling in the names of relatives, or devising strategies to circumvent the act using provisions like exemptions to plantations (Nair and Menon 2006, Radhakrishnan 1982). Many of the larger tenants also resorted to such measures. Evasion of the ceiling by tenants took place as the government first granted ownership rights to the tenants irrespective of the amount of land cultivated by them, and only then did it confiscate land in excess of the ceiling.[20] The inordinate delay in the implementation of the land reform legislation after the initial conception also gave plenty of room to the landlords and larger tenants to resort to such evasive tactics. The failure of the government to take over surplus land is illustrated in the steep fall in the availability of surplus land. While surplus lands were estimated at 708225 hectares in 1957, it fell to 44517 hectares in 1964 and further down to 40470 hectares after 1970 (Radhakrishnan 1981).

There has therefore been widespread consensus regarding the fact that the land reforms through the abolition of tenancy made land owners out of the intermediaries, but did not achieve the objective of giving land to the actual tiller of the land (Herring 1980, Oommen 1971, Mencher 1980, Govt. Rice Commission 1999), a fact that has been officially recognised as well (SPB 1997: 17). Kerala's land reform programme has been critiqued for having accorded rights in land 'too liberally' to families that had other primary sources of income, and who engaged in agriculture in a perfunctory supervisory manner (Jannuzi 1994: 138, Radhakrishnan 1982). Tenancy reforms have also been critiqued for not making a distinction among size categories, assuming that all tenants were equally deserving (Herring 1990). The largest tenant-cultivators of the pre reform era are therefore the biggest landowners today (Radhakrishnan 1981). It has therefore been argued that on equity grounds, a lower ceiling limit should have been imposed on large tenants, in the absence of which considerable concentration of benefits took place in the larger size categories of tenants (ibid.).

The benefits to the landless labourers were largely confined to the acquisition of ownership over homesteads (kudiyirippu) alone. Radhakrishnan argues that the land received by this group would pale into insignificance when compared with the land received by the cultivating tenants. The land was too small to even put up a hut, let alone for agricultural use (Radhakrishnan 1981). It is therefore argued that if land is viewed as a socially constituted bundle of rights, the Kerala reforms transferred the entire bundle that we call ownership to the tenants (Herring 1988). Further, by ignoring the existing divide between the large and small tenants, and in particular the landless labourers, the land reforms have been argued to widen the existing inequality in land ownership, rather than to reduce it (Radhakrishnan 1981). The inadequacy of the land reforms in ensuring a fair distribution of land is manifest in the re emergence of tenancy in the state in the form of lease farming, particularly amongst the agricultural labourers and sub marginal farmers (Nair and Menon 2006). It is therefore argued that the impression of land reforms having succeeded in the state is largely due to the fact that several intermediate and lower castes in the state came into possession of land, resulting in some socio economic mobility among them and not due to any radical re allocation of land holdings (ibid.).

Changing access to water

As mentioned earlier, the crop producing potential of a particular piece of land in the area is determined by access to ponds, streams and wells, as well as by its location in the catchment which determines its moisture retention capacity. Both these factors were neglected while reorganising land rights in the region through the land reform initiative. As a result, not all landholdings had equal access to water. The resultant inequity that was precipitated severely undermined the element of distributive equity that was inherent in the land reform initiative.

Those tenants who cultivated land in the ayacut of ponds, were given rights to the water in the pond as well. Those whose lands were not located in a pond ayacut therefore got no rights to water. The government made no effort in ensuring that land irrigated by ponds was equitably distributed. As a result, the large tenants and landowners retained lands that had access to water stored in ponds. The pond soon came to be viewed as the private property of those who owned land in its ayacut.

The following tables indicate the extent of inequity in access to water. The data presented in these tables is derived from the survey conducted amongst farmers. The data was collected from farmers belonging to three main landholding categories (large, medium and small) in the study area. For the present study, large, medium and small farmers were classified as those holding land above 2 ha, between 0.4 and 2 ha, and below 0.4 ha respectively. As mentioned in Chapter 1, a total of 144 farmers were surveyed from the 16 padashekhara samitis, with 48 farmers from each landholding category. In order to illustrate differentiated access to water amongst these three groups, percentages have been calculated for each group separately. While the total number of farmers interviewed is 144, percentages have been calculated separately for each of the three 48-member landholding group. Hence when I say that 89% of large farmers have access to ponds, it is implied that 89% of the 48 farmers from the large land holding group have access to ponds.

TABLE 4.2 Percentage of large, medium and small farmers with access to ponds

Land Holding Category	*Access to Ponds*
Large	89%
Medium	48%
Small	23%

Source: Field Survey 2002

As compared to 89% of large farmers, only 48% and 23% of the medium and small farmers respectively have access to ponds. This inequity is further aggravated by the fact that 37% of the large farmers have access to more than one pond. The corresponding figure for medium and small farmers is 8% and 6% respectively. The importance of access to water is reflected in the variations in land prices in accordance with access to water. Appunni and Harindran purchased land in different parts of the Velanganpadam padashekharam around the same time in 1996. While Appunni paid Rs 0.2 million per hectare, Harindran paid only Rs 0. 14 million per hectare. Appunni attributes the difference to the fact that his land carried water rights to the Velangapadam kulam, and was located immediately below the field channel supplying canal water. The land that Harindran bought did not fall in the ayacut of a pond, neither does he get adequate supplies of canal water.

TABLE 4.3 Percentage of large, medium and small farmers with access to one pond and more

Land Holding Category	*Access to one pond*	*Access to two ponds*	*Access to three ponds*	*Access to four ponds*	*Total Access to ponds*
Large	52%	29%	6%	2%	89%
Medium	40%	4%	4%	–	48%
Small	17%	4%	2%	–	23%

Source: Field Survey 2002

Further more, a high percentage of large farmers (67%) have access to ponds, which are individually, or family owned, while only 10% and 2% of the medium and small farmers do so. Being family or individually owned, the owners enjoy greater flexibility in taking water for irrigation. When an extended family owns the pond, the chances of disputes amongst farmers are reduced as well.

TABLE 4.4 Percentage of large, medium and small farmers with access to individually or family owned ponds

Land Holding Category	*Access to Individual or Family owned ponds*
Large	67%
Medium	10%
Small	2%

Source: Field Survey 2002.

Apart from access to ponds, small pits (kuzhis) and wells, the increasing spread of tubewells has also compounded the inequities between farmers of the different landholding groups. Owners of riparian land are also at an advantage when compared to owners of non-riparian land. The cumulative impact is manifest in the fact that 58% of the large farmers enjoy access to more than one privately owned source of irrigation. This implies that they have access to a pond and a tubewell, or a pond and a pit, or a pond and a well. In comparison, only 8% of the medium farmers and none of the small farmers have access to two sources of irrigation. Amongst this 58% of large farmers, 21% have access to three private sources of irrigation. This shows the range of access to water that large farmers enjoy, by virtue of owning land that falls in the ayacut of a pond, or land on which they have dug wells or tubewells.

TABLE 4.5 Percentage of large, medium and small farmers with access to more than one private source of irrigation

Land Holding Category	*Access to two private sources of irrigation*	*Access to three private sources of irrigation*
Large	58%	21%
Medium	8%	–
Small	–	–

Source: Field Survey 2002

TABLE 4.6 Differentiated access to irrigation

Land Holding Category	*Access to Ponds*	*Access to Kuzhi*	*Access to Wells*	*Access to Tubewells*	*Access to Stream*
Large	89%	40%	10%	35%	29%
Medium	48%	8%	10%	4%	23%
Small	23%	–	4%	–	8%

Source: Field Survey 2002

In contrast, a high percentage of small farmers (67%) do not enjoy access to any private source of irrigation. Their access to irrigation is confined to the water supply through the government owned canal system.

TABLE 4.7 Percentage of large, medium and small farmers without access to any private source of irrigation

Land Holding Category	*Percentage without access to any alternative source of irrigation*
Large	2%
Medium	19%
Small	67%

Source: Field Survey 2002

The issue of access to water was not considered while distributing excess land among the landless either, who mostly constituted the Scheduled Caste community. In other words, when landlords and large tenants had to surrender land in excess of the ceiling of 2.5 hectares per person, no stipulations were laid on the quality of the land that was surrendered. They were free to surrender the least wanted lands. In most cases, these lands were located in the upper slopes of a catchment, with poor water retention capacities, referred to as '*metu kandam*' or '*mele kandam*'.[21] Excess land (meaning land that is in excess of the stipulated ceiling on landholdings), referred to as *michabhoomi*[22] therefore primarily consisted of parambu or potta lands. Amongst the recipients of excess land interviewed during the course of the present study, almost all of them were given potta lands ranging between 0.12–0.2 ha, with very few owning more than 0.2 ha of land. In all cases, none of the excess landholders have been given water rights to the pond, even if the land acquired by them was located in the ayacut of a pond. The difficulty of raising paddy on these lands has led many of the excess land recipients to sell their land, thereby defeating one of the most important objectives of the land reforms. Paraman, himself a recipient of 23 cents of paddy land in Kallatukolambu says that out of the ten people given excess land in the area, only two of them have retained it.[23] Most of the michabhoomi lands located on the higher slopes were ideal housing sites as a result of which there was a high demand for such plots in the land market. At the other extreme, former landlords and most of the large tenants retained a substantial portion of the fertile kalayi lands. In an attempt to capture caste based inequity in access to water in the country, Tiwary argues that there is a need to extend the arguments of explaining inequalities in rural India from ownership of land to land use.[24] He notes that while scheduled caste communities have been given land, it becomes ineffective in the absence of timely and adequate availability of water. He argues therefore that there is a need to analyse what rural populations, who constitute the lower rungs of society (mostly from so called lower caste groups), do with the land they own, as it has 'wide implications for income, livelihood, and more importantly, capacities to break the poverty trap.' (Tiwary 2006).

Hence while the size of the landholding is one measure of equity in land rights, access to water is an equally important determinant of equity. This inequity is manifest in the fact that those who have access to water sources such as ponds, wells and kuzhis are able to stave off crop failure more effectively than those who do not. See Table 6.3 in Chapter 6 and the discussion that follows. This was further aggravated by a growing privatisation of rights to water, facilitated by the availability of water extraction technologies such as the pump sets. In the process, the large farmers who could invest more in such technologies could take the benefit of pumping at will from the water sources that were enclosed within their private land holdings.

What can be concluded from the land reform exercise in the study area is that conferring land titles to individuals without considering the distribution of water in a landscape that demonstrates considerable variations with regard to access to water can result in serious inequities. As Runge put it, 'the distribution of basic natural resources such as soil or water (including rainfall) is often random over time and space, and therefore assignment of exclusive use rights to a given land area can lead to an inherently unfair distribution' (Runge 1986: 623). The neglect of water rights while redistributing land has been reported from other parts of the country (Shanmugaratnam 1990) and the world as well. Abeyratne while examining the nature of property rights in land and water in Sri Lanka observes that though legislation and other interventions in the realm of land rights has had far reaching consequences for water rights, the latter ('water tenure') has been a subject of serious enquiry only recently (Abeyratne 1990: 73). In the Indian context, Tiwary has argued that landownership as the dominant mode of explanation for rural inequality has been carried too far, ignoring other modes of resource access, particularly access to water (Tiwary 2006). Of land reform programmes in general, Hodgson states that 'the concerns of water rights reform, scarcity and sustainability, are quite absent from the land reform debate' (Hodgson 2004).

The final outcome

By narrating the cases of a few large and small farmers in the area, I illustrate in this section the resulting inequities in land holding and access to water.

The large farmers today

Kochukuttan owns about 5.2 ha of land inherited from his father in the Manalipadam area. He has been cultivating this land for the past 50 years. Prior to the implementation of land reforms, his father had leased out this land from the Nedgungadi family. In addition, Kochukuttan had also fixed a deal for the purchase of an additional one-hectare of paddy land in the adjoining valley. Pandalamkulam (a pond covering an area of 0.96 ha) supplies water to his 5.2 ha land holding. This apart, water supplied through the canal network is brought to his fields by a field channel which ends in his field, from where it has to pass, field by field, to the lands of adjoining farmers. He also stocks canal water in Pandalamkulam, which insures him against crop failure. The additional one-hectare he proposed to buy encloses a perennial water pit as well. On parambu land he has raised coconut plantations, which are irrigated by a tubewell he dug during the early 1990s.

He lives in what was earlier the kalam, now converted into a fairly big house. His son is in the business of fish breeding.[25] Pandalamkulam is therefore used for both fish breeding and agriculture, as the entire pond and ayacut belongs to Kochukuttan. If other farmers had a right to the water in the pond, it would not have been possible to use it for fish breeding, for the demands for irrigation would have come into conflict with the demands of fish breeding. When the water level in the pond dips, they pump water from their tubewell into the pond.

Though details will definitely vary, Kochukuttan represents to a large extent the larger farmer in the area, owning relatively larger holdings and having access to multiple sources of water. He also represents the hitherto large tenant class, whose economic position underwent a significant transformation after being conferred ownership over the land cultivated by them. The most significant outcome was that they were no longer liable to payment of rent to the landlords. Additional amounts of water brought through the canal network during the same period, along with the introduction of chemical fertilisers and hybrid seeds, is reported to have led to a sudden spurt in yields during this period. The cumulative impact was reflected in a sudden spurt in their economic position.

Many of the tenants from this category re built their houses following the land reforms. In many a case, the klam over which they acquired ownership in the process of land reforms, was renovated. These kalams, which were named after the family names of the janmis such as the Ambat kalam or Parassery kalam, continue to be referred to by the same names. Most kalams by virtue of their position enjoy a scenic view of the surrounding paddy fields. The conduct of many a new land owner (a hitherto tenant), sitting in the kalam, giving orders to the agricultural labourers has often made one wonder: 'Is the janmi still around?.' One gets the same impression while talking to Velayudhan, who owns about 4 hectares of land in Peringotukavu. His father had leased out large stretches of land prior to the land reforms, most of which were divided amongst his children. Velayudhan has access to one pond, apart from water through the canal network. He also owns two rice mills in the area, which is an indication of his economic standing.Being located in the tail end of the canal, he has had to put up with uncertain water supplies since the 1990s.

The case of Bhaskaran, who is the president of one of the padashekhara samitis in the study area, is yet another example of the relatively privileged position of the well to do farmers in the area. He owns 8 ha of land in the area, which is irrigated by 3 ponds. Of the 3 ponds, he has exclusive rights to the use of two. A perennial pit, which has been deepened and widened, resembling a square well is also located in the low-lying fields in his possession. Of the three ponds, Maduteeni potta located higher above dries up very fast. Hence he pumps water from the well below into this pond so as to irrigate the plantains and vegetables he cultivates in addition to paddy. Being located in the tail end of the canal system, he has had to face crop failure in the past. Hence, he has now invested about 2 lakh (0.2 million) rupees in the sinking of two tubewells (240 and 270 ft deep) and pipelines, which bring in water pumped from the tubewells into his ponds. By pumping the water from the tubewells and shallow well into the ponds and keeping them full, he hopes to build a fool proof system, so that he is not affected by unreliable water supply through the canal network. Though the sinking of tubewells is becoming common, Bhaskaran's experiment with two tubewells and the elaborate pumping system was much talked about. Bhaskaran's house is one of the largest in the area. Being the president of the padashekhara samity, he is an influential person in the area as well.

Not all the large farmers in the area were tenants in the past. The families of some of the present day large farmers possessed land-owning rights prior to the reforms. This section of people usually comprised of high caste families, particularly, Brahmin and Nair families. Harihara Iyer is one such example. His family owned land in five locations, which were being managed by intermediary tenants (kanakkar) on their behalf. Anticipating the implementation of the land reforms, he quit his job in Madhya Pradesh and came back home to directly manage the family's agricultural lands, and thereby prevented their takeover by the tenants. His grandfather had set up the first rice mill in the area. The land that the family owned was divided amongst the heirs, of which Harihara Iyer now holds 4 ha, along with two ponds.

The small farmers today

At the other end of the spectrum, one finds the struggles of the small and marginal farmers, most of them owning less than fifty cents of land (0.2 ha). They include agricultural labourers, who have either purchased small parcels, or have received the 'excess land' taken over from the earlier landlords. They also include people engaged in a variety of occupations, which generate limited incomes with which they have purchased some paddy land. These include people working as toddy tappers,[26] as helpers in local hotels, driving bullock carts, running small grocery stores or home based enterprises such as tailoring units. Some of them have purchased paddy land, while others cultivated paddy on land taken on lease. The paddy grown on this land is mostly used for home consumption. It is the former however who comprise the majority amongst the small farmers.

As discussed earlier, amongst those farmers who have received 'excess land' from the government following the reforms, none of them have received water retaining kalayi fields. In many cases, sloping parambu land has been given on which it is difficult to grow paddy. Take the case of Velan, an agricultural labourer who works as a permanent agricultural labourer for the lands of Haridas. He was given 85 cents (0.34 ha) of parambu land in Vadukol in what is now the Kallatukolambu padashekhara samity. On this 85 cents of sloping land located at the top of the Vadukol hillock, two houses have been built, one for himself, and one for his elder son. Velan has levelled the remaining portion into four paddy fields, through manual labour alone. At the lowest point, he has dug a well and purchased an oil engine (one and a half years ago) for which he received some financial assistance from the panchayat. Apart from paddy he grows some vegetables on a small patch close to the well. Being parambu land originally, it does not receive water from the canal network, nor does it retain significant amounts of moisture. With the well he can now irrigate on and off, but it cannot sustain a crop of paddy, especially during the second crop period. Standing by his fields, looking towards the northeast, one can see the kalam and pond that belongs to Unniappa Menon's family. This family was one of the large land owning families in the region, and continues to own considerable amount of paddy land in the area. Looking at the paddy flourishing in the valley below, Velan's plight raises the question as to what the land reforms have achieved. Having had to invest almost all of his meagre savings and considerable physical labour, he still cannot be assured of a good crop of paddy. His question as to why the government gives such land as *michabhoomi* (excess land) finds no answers. Velan's desperation increases as days go by, having to raise seven girls and two boys, which includes a mentally retarded child. He raises a couple of cows, the sale of milk giving him some income to meet daily expenses.

Velan is only one among the many who face a similar predicament. Seventy-year-old Kunjan and his wife for instance, have worked as agricultural labourers, his wife continuing to do so. They belong to the Cherumakkal caste, and were brought here from near by Koduvayur by the janmis to work on their land during Kunjan's grandfather's time. They acquired ownership over their homestead following the reforms. They purchased 80 cents (0.32 ha) of paddy land located in the low-lying stretch of fields at Chatanchira. While this gives them paddy for home consumption, their sons eke out a living by going for fish business in the neighbouring district of Thrissur. One of their daughters in laws goes to work in the brick kiln as well.

Paraman in Kallatukulambu was given 58 cents (0.23 ha) of land as *michabhoomi* of which only 23 cents (0.092 ha) is paddy land. His house stands on the remaining portion. Paraman has not however been given water rights despite his land being located in the ayacut of the Kateri. So each time he wants to pump water from the pond, he has to pay for it, apart from the cost of renting the oil engine and cost of the diesel. Paraman works as a *karyasthan* (supervisor)[27] on the land of Ismail, an absentee landlord, for which he is paid about Rupees 2–3000 annually. His wife too is an agricultural worker. Unable to make both ends meet, agricultural labourers like Paraman and his wife go in search of other options to supplement their income. They have been going to the neighbouring district of Thrissur to work in the brick kilns for the past six or seven years. Paid at the rate of Rs. 250 for every 1000 bricks made, they work from four in the morning till eleven in the night, living in temporary sheds during the summer months of January–April. By April they return so as to commence work for the first crop of paddy.

Amongst the marginal farmers, single women, have an even more difficult time. Talking to Madhavi, at the end of her day's work, tells much of their plight. Madhavi, a widowed mother of six children, cultivates the 85 cents (0.34 ha) of land purchased by her husband from a large landowner about twenty years ago, with his savings earned from toddy tapping. They were however not given water rights in the pond while purchasing the land. They bought the land nevertheless, assuming that the water supplied through the canals would help them raise a crop. Today, Madhavi has to rely on Sunil who has one of the bigger shares in the pond ayacut, for water. Unable to work on her field and take care of the cows, she sold off the latter. Life is a continuous cycle of taking and repaying loans she says. She normally resorts to borrowing money from the rice mill owner to whom she sells her grain after harvest. The entire family participating in the agricultural work is a typical feature of most small farmer families. As much as possible they avoid hiring labourers, resorting to it only when absolutely necessary. Madhavi for instance hires labourers only during the time of weeding, for she alone cannot do it, and her children are too small to assist her.

As the above cases reveal, the inequality is not only in terms of extent of land holding, but also in terms of access to water. For many small farmers like Madhavi or Velan, it is the inaccessibility to water that makes agricultural unviable. This is made clear by small farmer Madhavi's dependence on the on the goodwill of the neighbouring farmer for water, who owns most of the land in the pond ayacut as compared to the control that large farmers like Kochukuttan or Bhaskaran exercise over water. Bhaskaran suffers a crop loss only because he is located in the tail reaches of the canal ayacut, a disadvantage he is able to somewhat negate by pumping water from his tubewells and ponds. Even if large farmers face crop loss, it affects only part of their total cropped area. In the case of small farmers, it implies no grain for the season at all. For a person like Madhavi for instance it means that she will have to repay the rice mill owner from whom she takes loans, without getting any grain at the end of the season. Many a time she says, the cows come and graze on the dry crop, while she repays the loan.

Fragmented Land and Fragmented Water

Fragmentation of land holdings has been an inevitable outcome of land reform programmes across the world. In the particular ecological context of Kerala where paddy is mostly grown in the valleys, effective water control requires that each valley or micro catchment be treated as an integral unit. In the process of conferring individual titles to paddy lands, the hydrological interdependence between paddy fields was ignored. Paddy fields however do not exist in isolation, being dependent on each other for the regulation of water supply and moisture regime. It is surprising that the high degree of interdependence between owners of separate land parcels was not given any attention, particularly when paddy fields were irrigated by a single source such as a pond or a canal. No attempt was made to spell out issues related to water management, an issue of importance given the increase in the number of individual land operators.

The fragmented approach to land and water management is well illustrated in the splitting of the pond ayacut into individual land parcels. In the pre reform scenario, each ayacut would be managed by 2–3 tenants at the most, with there being many cases where a single tenant cultivated the entire pond ayacut. When the first lot of tenants turned owners partitioned their land, the number of individual parcels within an ayacut increased. The absence of a mechanism to measure and monitor withdrawal rates of water has led to significant arbitrariness in the use of water stored in ponds. Even in cases where the entire ayacut was managed by siblings, the increasing nuclearisation of joint families (see p. 118 in Chapter 3) has led to the loosening of familial bonds. This had implications as far as joint agricultural operations were considered. In some cases, conflicts between siblings over partitioning of the family property has led to drawn out court struggles (see Box 4.2).

Apart from partitioning of the land, the rapidly growing land market led to frequent sales of independent shares as well. This is particularly so as many owners went to far away towns and cities in search of employment. In effect, the ayacut of a pond may be split up into numerous individual parcels, each one having a right to the water stored in the pond. Take the case of the Karinkulam, where there are 15 farmers who have a stake in the water, each one of whom is entitled to get water in proportion to their land holding.

Box 4.2 Court Battle Over Water

Ganeshan owns 3.2 ha in the Chittepadam area. When the land was managed by his grandfather, 0.8 ha plot was used as a *chera'* (low lying paddy land with high bunds, cultivated during the first crop season, and stored with water during the water scarce second crop period (see Chapter 2 for further details). When the land was partitioned amongst Ganeshan's father and uncle, his uncle was given the share that included the chera. With the introduction of the canals, his uncle converted the chera into paddy land (assuming that water would be available in plenty, doing away with the need of stocking water on the chera). Land lying downstream of the chera belonged to Ganeshan's father, who then went to court to contest the conversion of the chera, as his lands were deprived of the water that would be stored in it. The court case finally ended in an out of court settlement. The most important factor that came in the way of resolving the conflict was that there is no official dictum which states with clarity whether a chera can be converted into a paddy field or not during the second crop season. A similar conflict has been reported over the conversion of the Tumbikod chera nearby.

(Field Notes, 6.1.2002).

The fragmented approach is also reflected in the fact that the kalayi-potta distinction was not adhered to in the process of land redistribution. Kalayi-potta was not just a system of land classification, but was also a system of allocating water rights. This classification viewed land and water use in an integrated manner.

Confining double cropping of paddy to well watered kalayi lands and restricting water rights to potta lands were instances of an integrated approach. As per this system, all kalayi lands had access to water in the pond for two crops of paddy, whereas potta lands did not. If at all they did, it was only to the water that flowed through the higher sluice (mele ovu) of the pond. In some cases, lifting of water was permitted, but since manual lifting was the norm, it was resorted to only in cases of great urgency. Confining double cropping of paddy to kalayi lands helped to minimise the incidence of crop loss and to maximise the chances of a good crop on kalayi lands.

Along with the equity implications of ignoring the kalayi-potta classification, this omission had important implications on the discretionary use of water stored in the pond. In the pre reform situation wherein one or few tenants managed a single pond ayacut, cultivation of potta lands was often abandoned in order to save up the water for the kalayi fields where crops had a better chance of survival. Even amongst the kalayi lands, only the most low lying would be cultivated with a second crop of paddy when water supply was severely limited. In the years following the land reforms and the further fragmentation of individual parcels through partitioning and sale, each pond ayacut was split up into a large number of individual parcels, and each parcel consists of either potta or kalayi lands. Coordinated use of water to both these different land categories became increasingly difficult. The owner of potta lands was equally determined to raise a crop, despite the reduced availability of water in the pond. In a situation when a farmer owns only potta lands, asking him to forego cultivation on those lands is not possible, despite the lack of wisdom in cultivating a water intensive crop like paddy on potta lands. Discretionary use of pond water would have been possible only if each farmer's land holdings were dispersed in the ayacut area of the pond, such that each farmer owned both kalayi and potta lands.

Coordinated use of water within a pond ayacut has become increasingly difficult with the splitting up of the ayacut into smaller shares. Manikanthan recalls, that when certain fields in the ayacut of Veliya Eri needed water they would lift the water to one field and then dig a channel through the fields. The water therefore was not made to spread over all the fields on the way, saving on precious water. Once again this is not possible when there are many owners in a single ayacut. Farmers also recall that in the past, an agricultural worker would be assigned with the specific duty of ensuring that all the fields were getting adequate water. First thing in the morning, either the tenant or the labourer, as the case may be, would inspect the fields and assess which fields needed water the most. Today, the individual farmer will look at his parcel alone and come back. Given the decreasing size of individual land parcels in the pond ayacut, the interest of each landowner is confined to only a very small part of the pond ayacut. It would not be wholly incorrect then to say that farmers in the past managed the ayacut, whereas farmers today manage only their fields.

Individual parcels in the pond ayacut seem like an anachronism when viewed against the intended objective of the pond, that of storing rainwater and distributing it along the drainage lines of the catchment.

Fragmented ownership had an impact on activities related to pond maintenance as well. By altering land relations that were based on the janmi-kudiyan relationship, land reforms in the area brought about changes in the institutional mechanisms that governed the management of land and water in the area. In the specific case of water management, ponds comprised the most important irrigation infrastructure, which required periodic maintenance that was labour intensive. Farmers unanimously report that desilting has been rarely conducted in the post reform era. They attribute this trend to rising labour costs that would have to be incurred, and to the fragmented ownership of ponds. When more than one tenant became responsible for the upkeep of the pond, a feeling of shared responsibility towards pond maintenance was missing.[28] In the absence of an institutional mechanism that elicited the cooperation of the tenants involved in periodical repair works, regular pond maintenance suffered. It can be concluded therefore that while land reforms granted private land titles to cultivating tenants, it did not facilitate the formation of viable common property resource management institutions that ensured a sustained supply of irrigation water. Changing labour relations, discussed earlier in this chapter, have also played a role. Increasing wage rates (an outcome of the strong agricultural labour movement) made labour intensive activities such as pond desilting highly cost prohibitive for the new tenant turned landowners.[29]

The increasingly fragmented ownership patterns in pond ayacuts have given rise to a number of conflicts. These will be dealt with in detail in Chapter 6. Apart from the implications on equity, these conflicts also illustrate how a changed ownership regime impacts upon the management of land and water in a given area. The inability to view pond ayacuts as integral units has led to a situation wherein each pond ayacut has been fragmented into individual land parcels.

Land Reforms and Private Property: Implications on Equity and Ecological Sustainability

While land reforms abolished tenancy in the state, it also ushered in the era of private land rights on a significant scale. As in the case of many other Indian states, the 'creation of modern private landed property but with some degree of distributive justice' (Shanmugaratnam 1996: 169) was the main objective here too. In the process, the potential of alternative property regimes have been wholly ignored.

While correlations have been drawn between implementation of land reforms and agricultural productivity, the impact on land management and agricultural sustainability has received little attention (Mukhopadhyay 2005). Particularly so when land reforms have led to an entrenchment of private land rights. The neglect of the sustainability dimensions might be due to the fact that land reforms in general have been more preoccupied with the goal of socio economic equity through the redistribution of land holdings rather than ecological sustainability. In the whole debate over the intended versus the actual outcome of the land reforms in Kerala, the outcome of the reforms in terms of land and water management has received very little emphasis.

As a result, how the new owners were going to use their land was not an issue of concern at all. The increasing conversion of paddy land by private landowners is illustrative in this regard. Conversion of paddy land has played an important role in declining levels of paddy production in the state. Off late the state has witnessed a spurt in the number of cases where paddy land has been converted into housing plots and building sites, and Discussion with land brokers in the study area reveal that such a trend is intensifying in the area as well.[30] As of 2001, 1.41% of the paddy land in Kollengode panchayat (26.22 ha) and 6.19% (105.25 ha) of the paddy land in Elavenchery panchayat had been converted into plantations of coconuts, mixed crops as well as into building sites. There has also been an increase in clay mining from paddy fields for the purpose of brick making, which is in turn required to meet construction requirements.

PHOTO 4.1 Changing land use: brick making on paddy fields

Spatial land use planning and land use regulations have been imposed in many countries across the world, in order to restrain ecologically harmful land use practices. Kurnia for instance, suggests that if land use planning decisions are linked to procedures for obtaining water, governments would be able to impose land use patterns that are ecologically sustainable (Kurnia et al. 2000). Though their enforcement is weak in developing countries, they are not totally absent.[31] In Kerala, the Kerala Land Utilisation Order issued by the government in 1967 was the government's response to check the rampant conversion of paddy lands. By declaring illegal the conversion of paddy land, the act aimed at preventing further reduction in paddy production levels, as well as maintaining the unique ecological functions of paddy wetlands, given the important role they play in recharging the water table. Such land use regulations however are not 'inherent in the bundle of rights and obligations that make up land ownership' (Hodgson 2004: 24) as a result of which evasions do not lead to a confiscation of ownership rights.

Apart from the fact that private land ownership did not cater to the requirements of ecologically sustainable land management, sustainability was also threatened by the piecemeal approach of the land reforms and the nationalisation of private forests. The interconnections between parambu land, forestland and paddy land were ignored in the process of physical fragmentation. When parambu land was surrendered as excess land by the landed, and the same was redistributed to the landless, the functions played by parambu land in sustaining paddy cultivation (in terms of supplying green manure) were ignored. The same was the case when private forestlands were taken away from private management and vested with the state. Non-agricultural land was therefore not considered to be part of the extended resource base of farmers. The resultant reduced availability of green manure has now led the government to procure organic manure from outside the state and supply it to the farmers, in order to raise soil fertility. The implementation of land reforms and nationalisation of private forests therefore provide an illustration of the process of physically cutting up an agricultural cum forested landscape into private holdings and government parcels, neglecting the ecological interdependence between the two. Similarly, the fact that parambu land was more suited to agro forestry rather than paddy cultivation was not kept in mind while distributing the same amongst the landless. Hence while granting individual titles to land, the larger resource context that was crucial in ensuring agricultural sustainability was not given any consideration. While the need to protect the natural resource base of agriculture in order to enhance agricultural productivity has been raised in the Indian context (Nadkarni 1987: 364), in policy and practice agriculture continues to be viewed in isolation.

Conclusion

Overlooking the fact that land tenure provides the context and defines rights to water, the land reform exercise undermined the goal of equity to a large extent. While redistributing land, the focus was on the size of the landholding alone, and not on the quality of the land and the water rights associated with each plot of land. Coupled with the manipulative tendencies of the large tenants and landlords, it led to a situation wherein the landed retained the most fertile lands which also enjoyed access to water stored in ponds. The tying up of land rights with water rights led to a situation wherein those whose holdings enclose ponds and wells, and which border streams are buffered from crop losses. The conceptualisation of ponds and wells as the property of the concerned landowner led to a situation wherein the landed have access to private water sources, while the small farmers do not. The difference between large farmers like Bhaskaran and marginal farmers like Velan or Madhavi reveals the magnitude of the problem. This illustrates the fact that when ownership regimes do not match the natural variations of the resource base, equitable distribution is difficult to achieve.

Apart from the equity implications, this chapter has also illustrated the sustainability implications of the splitting up of land and water into private parcels. In the process of redistributing land, each pond ayacut has been cut up into individually owned parcels of land making integrated efforts at water management difficult. The grouping of hydrologically linked paddy fields into a natural composite unit, referred to as '*aela*' or 'padashekharam' has been taken up by the government time and again since the 1960s (see p.48 in Chapter 2). However, as the recently experimented with Group Farming programme revealed, fragmented and individualised land holdings came in the way of group efforts (Narayanan 2002).

The chapter has also illustrated the sustainability implications of the land reforms and the nationalisation of private forests. While it has been widely acknowledged that land reforms without reforms in broader support services such as access to markets, extension services, credit and infrastructure will not achieve the intended objectives of reform, it is only being slowly recognised that the focus on agriculture alone is too limiting (Herring 2000: 24). The need to view land not just as agricultural capital, but as a landscape that is a component of ecological systems (ibid.) is only slowly being recognized. Such a view would also imply that rights to agricultural land needs to be viewed in conjunction with rights to forest land and rights to water. Not only does this affect the sustainability of agriculture, but it also reflects the understanding of land as an ecological resource. Such an approach will have its implications while planning for any corrective or remedial measure intended to address resource degradation.

Notes

1 The critical point of difference between ownership and leasehold tenure is that in the case of the latter, there exists a separation of ownership from control. Under this form, while the ownership of land is vested with a particular person or institution, actual use of the land is undertaken by another. In this system the user of the land does not have permanent rights to land, the rights to the use of land being confined to the term of lease.

2 Though tenurial relations and social structures were extremely complex, high social standing has generally been associated with a right to income from the land without working on it, whereas those of the lowest social status have traditionally worked the land without owning it (Herring 1983).

3 'Malabar' written by Logan, popularly known as the 'Malabar Manual' was compiled during his tenure as Collector of Malabar. It is an exhaustive volume giving details of the geography, people, their religion and caste, language and culture.

4 The Board of Revenue for instance is reported to have maintained in 1818 that the janmi 'possessed a property in the soil more absolute than even that of the landlord in Europe' (Board of Revenue Proceedings cited in Menon 1994: 14).

5 Most of the personal accounts that one hears of the janmi-kudiyan relationship today (as narrated by farmers from their memories) pertain to the late colonial period, when janmis were in substantially powerful positions.

6 In 1916, in the landmark Olappamana case, the Madras High Court decreed that the ownership of non-tidal and non-navigable rivers was to be vested in the owners of banks on either side of the river' (Revenue Department G.O. dated 9 December 1929 cited in Menon, 1994:14)

7 'My father Thankappan, his father Teethu and his grand father Velappan, and now me and my brothers have been farming the same piece of land. When the reforms were implemented, my father was managing the cultivation here' (Field Notes 14.3.2002).

8 One such person by the name of Krishnan who used to work as a guard continues to be known as '*chungam* Krishnan'. Former tenants

recall paying a fixed amount to the guard while going to the forests to collect different varieties of leaves, which were dried and reserved as fodder for the goats during the monsoons, when it would not be easily available.

9 Young female agricultural workers, it is reported, would rather work as salesgirls and earn between Rupees 1000–1500 a month rather than stand in the slush and transplant paddy.

10 While the proportion of those engaged as agricultural labourers in the Kollengode panchayat declined from 52% in 1991 to 42% in 2001, those engaged in other occupations increased from 39% to 51% during the same period. The latter included those involved in livestock raising, fishing, mining and quarrying, manufacturing and processing, construction, trade and commerce. The change was even more significant in the Elavenchery panchayat. Here, the proportion of those engaged as agricultural labourers in the panchayat declined from 62% in 1991 to 47% in 2001, while those engaged in other occupations increased from 26% to 40% during the same period (Census of India 1991, 2001).

11 The implementation of land reforms in India was part of the overall objective of ensuring agricultural growth. The aim was to remove those impediments to enhanced agricultural production that arise from the agrarian structure, and in so doing to assure security to the tiller of the soil, as well as equality of status and opportunity, to all sections of the rural population (Third Five Year Plan in Balakrishnan 1999: 1272).

12 As per the Report on the Survey of Land Reforms in Kerala, 1966–67, 60% of the total households in the state had a holding size of less than 0.4 ha, comprising only 10.2% of the total area. In contrast, a mere 7.9% of the households had a holding size between 2–10 hectares in area covering 30.5% of the total area (Oommen 1971: 84). Holdings above 10 ha were largely plantations.

13 National Sample Survey 1962 in Oommen 1972: 53.

14 This purchase price was fixed as the aggregate of sixteen times the fair rent of the landholding plus the value of structures, wells and embankments of a permanent nature, and one half of the value of timber trees belonging to the landlord or intermediary subject to a maximum of sixteen times the fair rent.

15 The Act enabled the kudikidappukaran to purchase the land surrounding his or her hut up to 3 cents in corporations, 5 cents in minor municipalities and 10 cents in panchayat areas. When such a purchase is made, the land owner was entitled to only 25% of the market

value of the land, of which the kudikidappukaran needed to contribute only half, the balance being paid from the Kudikidappukaran Benefit Fund constituted by the Government (SPB 1997:4)

16 As per the stipulated ceiling, an adult unmarried person was allowed to hold only 5 standard acres (2 ha) subject to a maximum of 7.5 ordinary acres (3 ha), and a family of five members consisting of a husband, wife and their unmarried minor children could hold 10 standard acres (4 ha) subject to a maximum of 15 ordinary acres (6 ha). Standardisation was done on the basis of classification and productivity of the land, one acre of coconut garden or double crop wet land being considered as one standard acre. Exemption from the ceiling was conferred on essential items like lands cultivated with plantation crops like tea, coffee, rubber, cardamom, along with private forests, house sites and commercial sites (SPB 1997: 5).

17 Balakrishnan argues that no state in India, not even the communist regime in Kerala which initiated the process of land reforms in the state, 'has passed a land reform or agrarian relations act requiring the cultivators to till' (Balakrishnan 1999: 1272).

18 Sivaramakrishnan documents a similar process in Bihar where the time lag between the mooting of the nationalisation of forests programme and its final implementation enabled landlords, contractors, cultivating tenants and right holders to rapidly cut and lease out forests, defeating the very purpose of the Act (Sivaramakrishnan 1999).

19 Krishnankutty, a former tenant from the area recalls how his father who was cultivating on lease 26 ha of land in the hills, evaded the ceiling by giving a share to his married daughters, who were customarily not entitled to a share in the property. This share was however given to the daughters only on paper, with the land remaining under the ownership of his sons.

20 Radhakrishnan therefore argues that the land reform act should have conferred ownership rights on cultivating tenants only for the extent of leased-in area which was well within the ceiling limit, rather than conferring ownership for all of the leased in holdings and then acquiring the surplus lands (Radhakrishnan 1981).

21 '*Met kandamu*' and '*Mele kandam*' meaning high lying paddy fields.

22 Ironically, the word '*micham*' also has a derogatory connotation implying 'waste', or something that is not wanted.

23 Financial crises prompted many to mortgage their lands, which finally culminated in sale owing to the difficulty in paying back the borrowed amount.

24 Tiwari has assessed broad trends with regard to caste-based inequity in access to water in the country, by referring to existing secondary data, mostly from Census Reports.

25 As fish breeding yields rupees 40,000 in ten months, he feels it is economically more profitable than paddy cultivation, and therefore does not rule out a situation in the future when fish breeding would take over paddy.

26 Toddy tapping has largely been an occupation confined to the Ezhava caste group. Toddy tapping has proven to be a rather well paying occupation with the unionisation of toddy tappers. Some of the toddy tappers have been able to extend their land holdings through the purchase of additional land, and have built bigger houses as well.

27 The term 'karyasthan' is still in use despite pertaining to a bygone era. As mentioned earlier in the chapter, most of the land owning families did not even engage in the supervision of cultivation, and appointed karyasthans who would look into all the agricultural affairs of the janmi.

28 A similar outcome has been observed in a rather different situation by Sengupta is his analysis of traditional irrigation structures (the *ahars* and the *pynes*) in South Bihar (Sengupta 1996: 186). As a result of the break up of the land holdings of large landlords (*zamindars* as they were known) in south Bihar during the early 19th century, large estates and irrigation works which were earlier maintained by a single landlord were divided amongst several landlords. The poor state of irrigation works in the following period was attributed to this subdivision of landlords' interests, which led to a significant lack of cooperation between landlords with regard to the maintenance of irrigation infrastructure.

29 The present day alternative to manual desilting is the use of a JCB, hiring which costs about rupees 650–700 per hour. Once again only the very rich farmers can afford the same.

30 Land brokers in the area report a rising trend in the sale of potta and parambu lands, for construction purposes, particularly those located by the roadside. Potta and parambu lands are considered to be more suitable for construction purposes as they retain lesser amounts of water when compared to low-lying kalayi fields. Some of the potta fields require additional filling with sand before the basement of the building is constructed, for which sand would be mined from elsewhere and dumped here. Such a trend of paddy land conversion has been fuelled by the increasing risks of raising a second crop of paddy on these lands owing to the unreliability in the supply of canal water. It has

also been fuelled by the rising land prices. While low-lying paddy land was being sold at the rate of Rupees 0.3 million per hectare, parambu land was being sold at rupees 1.5 million per hectare. The increasing demand for housing sites also added to this trend. Those with money found it profitable to purchase land and then re-sell it in small plots of 0.012–0.02 ha.

31 The Indonesian government's response to the problem of rice land conversion in west Java was the introduction of spatial land use planning, wherein areas were divided into zones according to suitable land use. This measure was intended to prevent the conversion of fertile agricultural land (Kurnia et al. 2000). Such regulations on land use were prevalent in ancient custom as well. Goldsmith and Hildyard refer to the custom prevalent in the dry zone of Sri Lanka, which prohibited the construction of permanent buildings on prime agricultural land, with mud houses being the accepted norm for the majority of the population (Goldsmith and Hildyard 1984). Mud rather than brick houses were propagated as brick houses do not break when they collapse, while mud returns to the soil, providing valuable organic matter to the fields. In addition, the making of bricks entailed the digging of the earth up to a depth of several feet (ibid.).

5

Pockets of Scarcity

Distributional Maladies in the Canal System

'During the seven years since we purchased this land, we have been able to raise a second crop of paddy only on two occasions. Most of the time we lost the crop at the 'kadir' stage (the phase of panicle emergence). If we were assured of water at that point, we would not have lost the crop. If we had the power to mobilise large labour gangs at night, the water would have somehow reached us'
(Raju and Saraswati, owning sixty cents of land which is located at the tail end of the Vadakkemuri distributary at Poratancode).

'Those with access to ponds or those who own tubewells are able to save their crop despite the unreliability in the supply of canal water. Kochukuttan (a large farmer in Manalipadam) saved his crop by pumping for seven hours from the Pandalamkulam, after the third turn of canal water supply proved inadequate. Many like me, without access to ponds have repeatedly lost our second crop, getting only hay for the cows'
(Velayudhan, a small farmer from the tail end of the main canal).

This chapter examines in detail the actual distribution of water within the command area of the Chulliar reservoir. The Varayiri watershed is located in the second reach of command area of the Chulliar reservoir, which is a part of the Gayatripuzha Irrigation Project (see Figure 5.1). After a review of water control and delivery practices in the Chulliar command, the chapter shows how location in the irrigation system along with the operation of unequal power relations shape access to water. It illustrates that it is not always a shortfall in water that results in scarcity; malfunctioning infrastructure coupled with social conditions which impact on the distribution of existing supplies contribute towards the present crisis as well. The existing institutional mechanisms to address these problems are also reviewed.

The narration of water distribution is one that has been heard before from large canal irrigation systems in other parts of the country. One of damaged infrastructure, of corruption and misappropriation by head-enders. Essentially, the 'familiar combination of bureaucratic inefficiency and technical constraints' (Wood 1999: 779), aggravated by power plays along the canals, resulting in scarcity for some and abundance for others. I detail upon the distribution system in the Chulliar system for two reasons. One, it is to contest the oft cited allegation that a mere augmentation of supply would solve the problem of water scarcity. All head end farmers in particular argue that the tail-enders' woes would be resolved only if the existing storage in the Chulliar reservoir was increased. This has proved to be the biggest rationale for the proposing of augmentation schemes such as the Palakapandi project irrespective of its technological complexities and likely ecological consequences (see Chapter 3 for a discussion on this issue). While doing so, the existing distribution system within the Chulliar ayacut and its role in perpetuating scarcity has been neglected.

A second reason for describing the distribution system in detail is to shed light on the canal irrigation scenario in the district of Palakkad. As has been mentioned in Chapter 1, the district of Palakkad is considered to be the seat of canal irrigation in the state, with seven canal irrigation systems[1] irrigating various parts of the district. Water scarcity in Palakkad, particularly in the summer months has often been portrayed as a natural phenomenon, attributed to the relatively lesser amount of rainfall received in the area as well as to the drier climate. Of late, water scarcity in the eastern parts of the district, has been attributed to the non-compliance of the Tamil Nadu government with the laid out provisions of the PAP interstate water sharing agreement between the states of Kerala and Tamil Nadu (see Chapter 3). While all these factors do play a role, the role that a poorly functioning distribution network plays in precipitating wastage of limited supplies of water is overlooked.

The following box sets the scene for the water delivery problems in the Chulliar ayacut.

Box 5.1 Water does not reach me

Ramachandran's fields lie in Velampotta, a low lying area, located at the tail end of the Vadakkumuri Branch of the Kollengode distributary canal (see Figure 5.1). As per the distribution schedule, water is released to the Velampotta area for seven days. Ramachandran's fields are not directly served by the field channel that takes off from Vadakkumuri branch canal, which stops far short of his fields. Field channels are small water channels within the outlet command (usually irrigating about 40 hectares), delivering water to individual fields. He therefore has to wait for water to flow from field to field. Of the seven days when water is meant to run continuously through the field channels in this stretch, it takes 3 days for water to irrigate the fields and fill the pond located above his fields. Though the concerned landowners are aware of farmers further down waiting for their share, they continue with the practise of diverting canal water into their pond. Many a time, Ramachandran sends labourers (who work for him on a permanent basis) at night to open the sluices of the pond above, and divert the water into his fields. He also has to confront farmers located further down along the distributary canal, who block the outlet of the field channel that supplies water to his fields with sand filled sacks (shutters rarely being in place), so that water reaches them. By the time Ramachandran manages to get water to his field, the seven-day limit of water supply would have expired. According to the officials of the Irrigation Department however, water has been supplied to the Velampotta area for seven days, and hence his complaint of non-availability of water is not attended to.

The only way in which this seven-day limit can be extended is by appeasing the maestry who is officially entrusted with the task of supervising water distribution along the canals. Ramachandran leaves a hundred rupees at the Velampotta tea shop which is frequented by the maestry or gives to him personally when the latter visits him. When a number of farmers pressurise the maestry in this fashion, he tries to ensure that water supply to this section is extended for

> another couple of days. According to Ramachandran, he was able to raise a second crop of paddy during the years 1994–1998 only because the maestry 'cooperated'.
>
> (Field Notes 10.3.2002).

Ramachandran's case raises issues of pertinence to the existing distribution pattern. The existing system of water distribution in the canal system, the present state of art of the irrigation infrastructure as well as the present functioning of the irrigation department personnel are some of the key issues. A discussion on these issues necessitates an understanding of the intended system of water distribution in the canal system mediated by the regulatory functions of shutters, locks and keys, and the supervisory role of the irrigation department personnel, especially the *maestry* in supervising water distribution along the canals.

Water Allocation and Distribution in the Chulliar Ayacut

As mentioned in Chapter 2, the Left Bank Canal of the Meenkara reservoir links up with the Chulliar reservoir at 6.20 kms, and then continues as the Left Bank Canal of the combined system up to 20.5 kms. The intended ayacut area of both the reservoirs has undergone changes. Originally, the Meenkara reservoir was to supply water to 1012 hectares of land through the LBC,[2] and only the remaining 2428 hectares was allotted under the Chulliar ayacut. In practice however, only the area upto 6.2 kms, i.e. upto the feeding point from the Chulliar L.B. canal is irrigated by the Meenkara reservoir, enhancing the area to be irrigated by the Chulliar reservoir to 3074 hectares (GOK 1998).

The ayacut area of the Left Bank Canal is divided into the first and the second reach. The Varayiri watershed is located in the second reach. The shutter at Virutti along the main canal, which is the point of division between the first and the second reach (see Figure 5.1) is located at the southeastern corner of the watershed. Amongst a total of eight distributary canals in the command area of the Chulliar reservoir, four distributaries (Kollengode, Payilur, Karinkulam and

FIGURE 5.1 Chulliar Canal Network

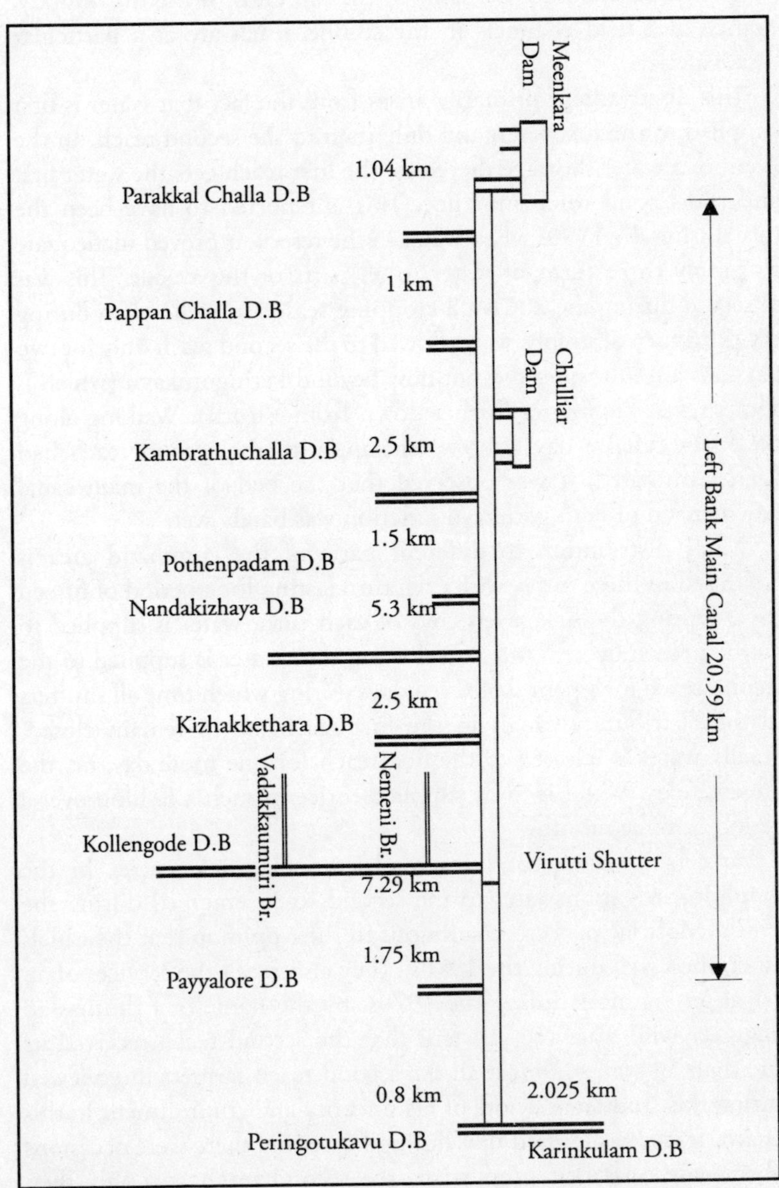

Source: GOK 1998

Peringotukavu distributaries) are located in the second reach. As the case of Ramachandran has shown, the tail ends of the distributary, branch and field channels in the second reach are at a particular disadvantage.

This disadvantage primarily arises from the fact that water is first supplied to the first reach and only then to the second reach. In the event of a water shortage therefore, the first reach gets the water first and the second reach may not. This is reported to have been the case during the 1990s when water in the reservoir proved inadequate to supply three turns of water to all parts of the ayacut. This was observed during the 2001–02 cropping season as well, when during the third turn of supply, water flowed to the second reach only for two days. As a result water did not flow beyond Peringotukavu (which is about seven kilometres further down from Virutti). Walking along the main canal a day after water supply to the second reach had been terminated, it was observed that the bed of the main canal downstream of Peringotukavu junction was barely wet.

Water distribution to different parts of the command area is organised in three turns, with each turn lasting for a period of fifteen days. During the first seven days of each turn, water is supplied to the first reach up to Virutti. Following this, water is supplied to the second reach for a period of seven days during which time all shutters to the distributary canals in the first reach are to remain closed. Finally water is released to the first reach for one more day, i.e. the fifteenth day. Water is to be supplied thrice in such a fashion over a period of three months.

Farmers in the second reach observe that deficiencies in the supply of irrigation water to the second reach emerged during the 1980s. Most farmers are unanimous in their opinion that the canals functioned well during the 1970s. They also recall the services of an irrigation engineer during the 1970s, by the name of Ezhuthassan Engineer who was very insistent that the second reach received its due share of water. Almost all the second reach farmers interviewed during this study made note of his sincerity and commitment in this regard. It is also reported that during his tenure there were occasions when water was first supplied to the second reach and only then to the first reach. Sankunni Master, an old school teacher distinctly recalls that water was supplied to the second reach first during the Emergency period[3] (i.e. during 1975–77).

The mid 1980s is reported as a turning point, which is when most parts of the state were reeling under a severe drought. During the mid 1980s, the deficiency in rainfall and the inadequate supply of water from the adjoining Chittoorpuzha system, from where water is diverted into the Chulliar reservoir (see Figure 3.1 in Chapter 3), is reported to have led to severe water scarcity in the Chulliar ayacut. Farmers recall that this was the time when most of the regulatory devices (outlets and shutters) were damaged by irate farmers, desperate to divert the water into their drying fields. The 1990s was a difficult decade for the second reach farmers, and it is reported that water has not flowed through certain sections of Valluvakundu-Tekinkadu, Peringotukavu, and Tumbidi-Karipayi padashekharams for five to eight years continuously. Farmers from areas beyond Karinkulam for instance have been exempted from the payment of the water tax for this reason. Location therefore plays an important role in access to water. Mollinga uses the term 'natural queue' to describe such a situation, wherein a farmer's location along the canals predisposes him/her to more or less water (Mollinga 2003:202). This priority given to the first reach is aggravated by the poorly functioning regulatory devices along the main and distributary canals, as well as along the field channels in all parts of the ayacut. Hence even when water is being supplied to the second reach, the water flowing through the main canal up to Virutti leaks out through faulty or missing shutters to the first reach. This leads to a situation wherein many parts of the first reach continue to be supplied water when they should not.

The system of supplying water first to the first reach has been so well entrenched that even a suggestion to reverse this system is uniformly shot down by farmers in the first reach. In the course of a discussion on this matter with farmers from both the first and second reach, many farmers in the second reach opined that water should be first supplied to the second reach, at least in alternative years. They cited the case of it being done during the 1970s. Farmers in the first reach however shot down this suggestion saying it would be akin to serving food to a queue of hungry people, beginning with the last person in the queue. No hungry person would tolerate the sight of food displayed in front of them, without being able to eat it, they argued. The irrigation department officers were also not inclined for such a reversal. According to them, while they would be able to justify water scarcity in the tail ends of the second reach on account of the location factor, they would not be able to justify water scarcity in the first reach.

Unlined canals

The full length of the main canal had not been cemented when the study was conducted (i.e. during the 2001–02 period). This was the case with certain portions of the distributary and branch canals as well. In some stretches, the unlined portions having grassy beds have been considerably silted up. The branch canal that branches off from the Payilur distributary towards Tachakora for instance, is so heavily silted up and colonised by grass and weeds, that one has to be told that it is a canal. It resembles more of a shallow stream channel. Similar is the case with most portions of the Karinkulam distributary shown in photo 5.1.

While the Irrigation department is supposed to ensure regular maintenance and desilting of main and distributary canals, farmers report that the desilting is mostly done in a hurry, just a few days before water is released from the reservoir. During the 2001–2002 cropping season for instance, the desilting and maintenance work in the area beyond the Kachamkurishi Temple involved only the removal of major blockages in the canal which mostly resulted from dumping of waste in the canals. No desilting of the raised canal bed was actually done. In addition, the so-called desilting work was not completed before the water was released. As a result, water supply to areas beyond the Kachamkurishi Temple was further delayed. It needs to be noted that this is the tail end region of the main canal, where farmers sow earlier than the farmers in the head reaches in order to harvest the crop before the summer months. These farmers were therefore most needy of water (as the crop in those parts had attained a growth of 40–50 days) at that point of time.

Photo 5.1 The silted Karinkulam Distributary in the tail end of the second reach, through which water has not been supplied since the late 1990s owing to non-availability

Enlarged outlets

The most important control systems that regulate the distribution of water in the Chulliar systems are the outlets, sliding shutters, and the system of lock and key. The size of the outlet is critical, for it is adapted to the size of the command area behind it. This implies that with a given water depth in the main or distributary canal, the pipe outlet releases exactly that amount which is required to support a particular cropping pattern in the command area (Mollinga 2003: 149). Very few outlets however retain their original size, as they have been enlarged by farmers who have at some point, desperately attempted to divert the water into their fields. Subsequently, they have not been repaired. This is applicable to the size of the outlets along the main and distributary canals, as well as along the field channels. The four inch outlets which regulate the flow of water along field channels have all been enlarged into much larger ones.

Chandran from Matacode observes that during certain seasons, farmers would have been desperately waiting for the canal water to reach their fields. The paddy crop would be at the verge of wilting. When water is finally released, instead of patiently waiting for their turn, they enlarge the outlets so that additional water flows through their channel and reaches their fields faster. Nooruddin from Tahsildaarpadam remarks that water would reach their fields when the outlets were four inches in diameter. Now that they are much bigger, it is only the farmers at the head end of the channel who get water enough to raise a crop. Few others like him did not even raise a crop during the 2001–02 second crop season, knowing fully well that water would not reach them. Nooruddin says that only if the outlets retain their original dimension, and only if shutters are properly placed, will people like him get water.

Missing/leaking shutters

Like the outlets, the shutters also play an important role in regulating water flow. Apart from the shutters present at the head of the distributary and branch canals (the head regulators), the flow through the outlets along the distributaries and branch canals is also regulated by shutters. A total of forty-four outlets divert water from the distibutary and branch canals, into field channels in the second reach. Canals that take off from branch and distributary canals are referred to as minors or tertiary canals in certain states. In Kerala, the irrigation department refers to them as field channels. As noted in Chapter 2, the local people also refer to them as CADA 'chaals', as they were lined by the CADA department under the CAD programme that was implemented in the area during the late 1980s.

As per the rule, when water supply to a particular section is terminated, shutters should be lowered across the outlets and locked. Observations of water distribution along the Kollengode, Payilur and Peringotukavu distributaries and the field channels revealed that this system was not followed. Out of the forty-four outlets that supply water to field channels, in fifteen cases, shutters had been physically removed by the farmers in order to take water out of turn. Farmers from further down below have to come and block such outlets with sand bags and plug the sides with hay.[4] In some cases, the shutters had not been lowered even after water had been supplied for the scheduled period, and in some others while the shutters had been lowered, the rusted and damaged conditions allowed water to flow through them. Hence, whenever shutters had been pulled down to block water flow through the outlet, water was found to escape through the sides and the bottom. I observed three cases where shutters had been lowered and locked, but the key was in the hands of farmers near by and it was reported that they would raise the shutter during the night. The crux of the issue is that the flow through not a single outlet was being adequately regulated by the shutter. Wastage precipitated by leaking and missing shutters and by enlarged outlets is further aggravated when the canal traverses through high ground. The drop in height from the level of the main or distributary canal to the paddy fields below causes a larger amount of water to escape through unregulated outlets. The field channel diverting water from the Poolaparambu outlet along the main canal illustrates the above phenomenon. This field channel encounters two drops in height before reaching the paddy fields below. When the shutter to the outlet was fully raised, the force with which water flowed downwards was found to destroy the bunds of the paddy fields below. The concerned paddy farmer, who was found to lower the shutter in order to reduce the force of the flow, complained that farmers from further below would raise the shutter soon after he has lowered it. Even when the outlet is closed after water has been supplied to the field channel for the scheduled number of days, the damaged condition of the shutter causes water to continue flowing downwards. Saving this water would help some tail end farmer further down struggling to prevent his crop from drying.

The main shutters along the main canal, which regulate flow to the distributary canals, are also found to leak. The shutter at Virutti for instance, which regulates the flow of water to the second reach is one such case. Even when supply to the second reach has been terminated, some amount of water continues to leak from under the shutter. This '*leak vellam*' (*vellam* meaning water) as it is referred to, flows for at least a hundred metres, which is made use of by farmers in that stretch. The same has been observed in the case of the Perottukavu shutter, which regulates the distribution of water to the Vadakkemuri section along the Kollengode distributary canal.

A proposal for the revamping of the Gayatri Project[5] submitted by the Chittur Division of the State Irrigation Department in 1998 made note of the deterioration that had set in as far as the canal infrastructure and related control devices were concerned (GOK 1998). On the basis of a survey undertaken to assess the condition of the reservoirs and the conveyance system of the Gayatripuzha irrigation project, the report noted that the damage was widespread, extending from the operating platform of the spillway shutters of both the dams (the Meenkara and the Chulliar), to the earth and masonry dams, and also to the control systems like the cross regulators and shutters. With regard to the control systems specifically, the report observed that the damage was so severe that they could not be operated, with the shutters and related structures requiring either replacement or repairs (ibid., 64). According to the report, unlined canals and faulty control systems precipitated conveyance losses of a high order, calculated as 48.43% (GOK 1998). In the state as a whole, conveyance efficiency in completed projects has been estimated at only 60%, which has been considered as one of the factors that have led to an 8.62% reduction in the area irrigated by government canals during the period 1990 to 2002 (GOK 2003). A more recent survey conducted by the Irrigation Department in 2002 in the Chulliar command area, corroborated the above findings, noting that none of the shutters were in place.

The malfunctioning of the control systems leads to a situation wherein there is very little regulation of water flow. Hence, while the second reach may be supplied water for the scheduled seven days, the continuing appropriation of water by farmers in the first reach through enlarged outlets and missing shutters reduces the strength of the flow. This has severe implications as far as the tail ends of the main canal and the distributary canals in the second reach are concerned. Along the main canal for instance, water does not flow beyond the Karinkulam junction, from where the Karinkulam distributary takes off. Not a drop of water flowed down the two kilometre long Karinkulam distributary. While the tail enders do not get water even when water is being supplied to the second reach, the head enders continue to benefit from supply of water, even after they have taken water for the scheduled seven days.

Power and Location

In a detailed analysis of water distribution practices in the command area of the Tungabhadra Left Bank Canal in Karnataka, Mollinga identifies two factors which contribute to the unequal distribution of water: the earlier discussed location factor and the economic and political power of farmers amongst whom the water is distributed (Mollinga 2003: 202). The previous sections have dealt with the locational disadvantages of farmers in the second reach as well as the poor state of canal infrastructure and malfunctioning regulatory devices. The following sections shall focus on how power plays along the canals shelter the socially and economically powerful farmers from water scarcity. The intervening of power relations in the distribution of water is manifest in two ways – in the phenomenon of first reach farmers appropriating more than the amount due to them, and in the process by which the powerful farmers in the second reach, particularly in the tail ends of the second reach are able to mobilise water to their fields when the majority of small holdings dry for want of water.

The display of power along the canals

'Trying to get more than their authorized shares' (Ostrom 1990:6) captures the misappropriation tactics resorted to by farmers located in the head end of the main, distributary and field canals. This they do with the silent, but almost total support of the irrigation bureaucracy. Chambers' analysis of head end misappropriation in Indian irrigation systems in the late 1970s, describes how head enders indulge in practices such as 'constructing illegal outlets, breaking padlocks, drawing off water at night, and bribing, threatening or otherwise in some way inducing officials to issue more water' (Chambers

1977:355). Thirty years later, the description aptly sums up the situation in the Chulliar ayacut. Misappropriation of water in this case is made easier by the damaged distribution infrastructure, as has been detailed in the earlier sections.

The damaged conditions of shutters along with missing shutters at certain outlets makes it easy for farmers in the first reach to continue taking water out of turn. As has been observed with water distribution in many other large scale irrigation systems, farmers prefer to negate the regular rotation system, and go in for blocking and unblocking outlets, so as to get more than their authorised share (Ostrom 1992: 6). On two occasions, large farmers were seen biking along the main canal accompanied by agricultural labourers. On one such occasion, the large farmer instructed the labourer to plug certain outlets so as to ensure more water to his fields. Majority of the large farmers are found to employ labourers at night to close and open required outlets. Some of them are found to enjoy greater proximity with the officials of the irrigation department. One such farmer is known to host a party for the officials of the irrigation department before the commencement of each cropping season. He is also known to invite all the maestries to his home and to treat them to food and liquor. Small farmers from the area complain that blatant violations of the rule by him and his labourers go 'unnoticed' by the department. Of the two other large farmers who were observed to enjoy personal affinity with the staff of the irrigation department, one had a political background and the other had made his fortune as a contractor. Similarly, there have been instances when the lock and key of the shutter has been entrusted with a large farmer of the area. In one such case, the farmer had a holding of less than a hectare, but was a former elected representative of the panchayat. She explained the practice of the maestry leaving the lock and key of the shutter with her as follows: 'He normally does not leave this with all and sundry; he leaves it only with whom he trusts.'

Taking water out of turn is not followed by farmers in the first reach alone; it is also followed by the larger and more influential farmers in the second reach. Farmers in the second reach do not form a homogenous category. The existence of power plays amongst them is manifest in the fact that the large farmers from amongst them are able to somehow get some canal water into their fields, while the majority of the small farmers suffer crop failure for want of water. During the second turn of release in the 2001–02 second crop season in the month of January,[6] the District Collector had instructed that a maestry be stationed at Virutti when water was to be released to the second reach. The Collector's order followed an incident wherein a shutter in the first reach was slashed open by an impatient farmer during the first turn of release. Complaints lodged by the second reach farmers had prompted the Collector to take this step. While I was discussing this matter with the maestry posted at Virutti as per the Collector's order, a large farmer from Peringotukavu (a tail end area), came on a bike, and spoke to the maestry in a very haughty manner. He was ordering the maestry to close all the outlets in the first reach, and the maestry was replying in a very meek tone. When he left, the maestry told us that while the farmer's anger was understandable given the lesser priority accorded to the second reach, his efforts to take water to his fields would deprive others in the second reach. The next day, we observed that the water had reached the concerned large farmer's fields and he had filled one of his ponds too with canal water. Farmers immediately upstream however, were complaining that he had closed the outlets to the field channels that irrigated their fields.

Box 5.2 Conflicts between tail enders in the second reach

1. Narendran owns 1.6 hectares in the Velampotta area, his fields being located at the tail end of the branch that supplies water to the Vadakkemuri section. Though his holding is not small by the average land holding standards in the area, he is in a difficult economic situation. He says that when he along with farmers from the area go to open the outlet to the field channel that irrigates their area, they have to encounter the labour force hired by Govindan who owns land further down below. Govindan is a wealthy farmer who owns a textile shop in Kollengode town. He bribes the maestry, who then allows his workers to open and close shutters, so that water reaches his fields and fills up his ponds.

2. Farmers whose fields are irrigated by the Payilur distributary complain that when large farmers from the tail end of the main canal

get desperate for water, they first come and close the outlet to the Payilur distributary, while those below may continue to be open. The farmers located at the tail end of the Payilur distributary, whose fields may not have access to even a field channel pay the price for the same.

(Field Notes, 15.5.2002).

A similar case was observed from Kizakemuri, which falls at the tail end of the Peringotukavu distributary. Large farmers like Ajayan from the tail end area of Kizakemuri are reported not to have slept for an entire week, in order to somehow get the water to flow into the canals that irrigate his holding of 2.6 hectares. He would take a jeep load of labourers to close the outlets all along the main canal, even those located in the second reach, so that adequate water reached his fields. Another incident cited by many farmers and the maestry was that of water reaching a large farmer's pond in the second reach for a single night in the previous years second crop season (2000–01). It was reported that during the third turn of water supply when water was exclusively released to the first reach, the shutter releasing water to the second reach at Virutti was raised one night and by the next morning, the above mentioned pond located by the main canal was full of water. As Unni Nair from Poratancode stated, 'When water is being supplied through the canals, farmers do things that they cannot talk about in public'.

Each such act of misappropriation deprives some farmer of his share of water. As one moves towards the tail ends of the distributaries, the conflict is between tail enders, the more powerful amongst them getting more water. It needs to be noted however, that beyond Peringotukavu in the second reach, farmers do not take as much effort in closing and opening shutters and getting water into their fields. The reason they cite is that however much they try, they would not be able to get the requisite amount of water. They may get a trickle, but that won't be enough for raising a crop of paddy they say. Hence, many large farmers in this stretch have invested in private solutions through digging of wells and tubewells, and through pumping water from the stream. A few others abstain from raising a second crop. Ismail for instance who stays in Palakkad town would rather not raise a second crop than engage in 'night-time activities' which cause immense mental harassment to him. Another large farmer reported that even if they were to send labourers to do the job at night, many of them get drunk and pick up unnecessary brawls which may even culminate in the police filing a case against them. So apart from the hundred rupees that has to be paid to each labourer, one may have to plead for them at the police station.

The plight of women farmers

Amongst the small and less powerful farmers, the women farmers face an especially trying time. All the women farmers interviewed as part of the study were widows. They would not venture out at night lobbying for water. As sixty year old Karthiayani Amma who owns land in Pulikalpotta put it, "How can I guard the outlets at night?" Some of them would pay labourers to get them some water. Apart from labourers employed by specific landowners to stand guard at outlets and divert water to the desired location, there are others who will do the same for any farmer, on being paid a sum of Rupees 100–200. During the second crop season, many such labourers make their presence felt along the canals, asking the farmers "Do you want water for a night"?.

Madhavi, a widow and mother of six children (see p. 172 in Chapter 4 for a discussion on the plight of small farmers like Madhavi, despite the implementation of land reforms), cultivates a mere 85 cents (0.34 ha) of paddy land in Kizakemuri. She is never sure till the very end whether she will be able to raise a good crop of paddy. Each time she has to pay workers to get water into the field, as she has no one to go and do this job. This time too (2001–02 season) she paid labourers Rs 200 to get her water.

The disappearing maestry

Chambers argues that the over appropriation of water by head enders is difficult to control, unless institutional controls in the form of countervailing custom, social sanction or physical force prevents the same (Chambers 1977:349). The most important form of institutional control as far as day-to-day field level water distribution in the Chulliar ayacut is concerned is the supervisory role played by the maestry. He is entrusted with the task of ensuring the rotational schedule of water distribution, i.e. ensuring that gates to various distributary canals and field outlets are opened and closed as per schedule.

Farmers report that during the initial stages of the functioning of the reservoir and canal system (during the 1970s), the maestry discharged his duties diligently. The fact that all shutters were locked as per schedule is borne by the fact that farmers would have to personally request the maestry to raise a lowered shutter out of turn for a few hours, for which bribing was not uncommon. Rajan from Kannankolambu recalls an instance during the 1970s when he went and pulled out a maestry from the cinema hall, in order to raise the shutter for him. The presently damaged condition of the outlets and shutters has created a situation wherein the presence of the maestry is not required and the farmers themselves control the flow of water. Maestries today sit at one of the sub offices of the irrigation department at Payilur, and are rarely present on the canals. Farmers report that when the second reach farmers complained about the absence of the maestry to the District Collector, the latter rebuked the maestry and asked him to recall the past when they would be present on the canal bund, dressed in khaki trousers. Over a period of time, farmers have taken control over field distribution, and the maestry has retreated from active supervision, so much so that he is rarely present along the canals when water is being distributed today. During the daily observations of water distribution practices in the second crop season, I observed the maestry discharging his duties only on one occasion. Even this presence was a result of the order issued by the District Collector, in order to ensure that water reached the second reach.

Maestries were however reported to have been present at the local tea and toddy shops, and of even having visited some of the farmers at their homes. Farmers have also reported that they have bribed the maestries so that the shutters of outlets supplying water to their areas remain open. So while maestries are not found to be actively discharging their duties, they do make their presence felt in the area. The main reason that maestries cite for their increasing absence along the canals is their disinclination to get involved in brawls with farmers and their labour gangs.

Storing of canal water in ponds

An issue that has a close bearing on the distribution of canal water is its diversion into ponds. This takes place amidst two prevailing perceptions regarding the ownership of canal and pond water. While water that flows through the canal is viewed as state property, to which all farmers in the ayacut are entitled, water stored in the pond is viewed as the private property of those who own land in its ayacut. Since not all farmers have access to 5.2 ponds, the storing of canal water reduces the total amount of water that should be made available to all farmers in the ayacut. It also has a serious impact on the timeliness of water supply. Farmers in the tail ends in particular are severely affected as it increases their period of waiting.

The fact that all farmers go about filling their ponds with canal water in a very discreet manner indicates that it is not considered to be 'right'. Most farmers divert canal water into their ponds at night. Some farmers hire labourers especially for this purpose, who guard the concerned shutter. They are also found to bribe the maestry with money or liquor, so that he turns a blind eye to such diversions. Canal water is diverted into ponds in three ways: by openly diverting the water through paddy fields into ponds, by breaching field channels and diverting the water into ponds and by siphoning water from the canal into the pond (see photo 5.2). Breaching of field channels has been observed in three cases. Take the case of the Poricholam kulam (known as Poricholam Potta) which is located below the Ennayaram field channel, a short distance after it branches out from the Vadakkemuri distributary. The side of the field channel has been breached in such a way that when the field channel is in full flow, a significant amount of water runs down into the pond below. It is portrayed (by the owner of the pond) as yet another instance of poorly maintained field channels. Farmers from further down come and plug the leak, which is removed once they are gone. A similar example is observed in the Velanganpadam area. The Velanganpadam kulam is located at the head end of the field channel. A breach was observed along the side of the channel, keeping the Velanganpadam kulam full whenever water is being supplied. While many farmers from downstream complain bitterly, the irrigation department has not taken any action in both the above cases. In the case of the Velanganpadam kulam, farmers from downstream had filed a complaint with the irrigation department in 2002. A third case where this has been observed is the field channel at Illathu Padam that has been breached such that water flows into the Illathu Padam kulam. The siphon method is resorted to when the pond is located adjacent to the canal.

The case of the Kallankulam in Mannathupara stands out in this regard. The road in between the main canal and the pond prevents

PHOTO 5.2 Siphoning of water from the canal into the pond

the diversion of canal water into the pond. Hence, Chandran a farmer who owns land in the ayacut of the pond was found to siphon out the water from the canal. While doing so, the hose that was used to siphon the water was wrapped in green leaves to prevent easy detection. This was a permanent fixture throughout the second turn of supply to the second reach in February. This pipe was removed by other farmers, only to be replaced again. In the case of the Tayamkulam located by the main canal, three hoses were simultaneously siphoning water into the pond. Water sales from the Tayamkulam have been widely reported from the area, with the owners of the pond raising money in such a process. Similarly water siphoned from the main canal to the Dharmi temple pond has also been sold during periods of high demand for water. Water siphoning from the main canal at Mele Cheerani to the Mele Cheerani kulam enabled the pond owner to fill it with canal water on a single night during the 2000–01 second crop season. The most glaring manifestation of the inequity in filling ponds with canal water is when farmers sell the water so stored during times of peak scarcity. There have been instances of farmers storing canal water in ponds even when they have not cultivated a second crop, such that they can sell all the water.

Molle notes that while a farm pond can be viewed as a conservation measure if it captures some canal water that would be further lost, it may also lead to re appropriation if this water was ultimately to be used by some downstream user (Molle 2003). The issue of re appropriation has not been recognised by the Irrigation Department. In the earlier mentioned proposal for the revamping of the Gayatripuzha Project, the department notes that the water saved through improved canal infrastructure could be used to 'supply water to tanks to augment tank fed irrigation' (GOK 1998: 95). This seems a rather naive statement that is ignorant of the inherent public-private issue in the transferring of canal water into ponds. This pubic-private divide as far as the use of water is concerned shall be taken up for further analysis in Chapter 6.

Water distribution below the outlet

While damaged infrastructure precipitates inequitable water distribution above the outlet along the main and distributary canals, the situation below the outlet is not very different either. As mentioned earlier, field channels carry water from the outlets along the main and distributary canals to the paddy fields. Field channels however have not been constructed for their entire length, leaving a significant area to be irrigated by field-to-field irrigation system.

A common critique of irrigation projects across the state is that while efforts have been focused on the construction of medium and large-scale storages, a sufficient number of field channels to take the water to the fields have not been laid (CWRDM 1981). As in other parts of the country, field-to-field irrigation has been known to precipitate higher losses of water that can also result in over irrigation in fields near the outlets and under irrigation at the tail end reaches (SPB 1975: 45). Experiments with field–to-field irrigation indicate a water loss of 30% as compared to irrigation through field channels (Chackacherry and Jayakumar in Jose 1991: 89), as water has to spread through an entire field before it passes on to another. It does not reach its intended location at the end of a field channel, and hence disappoint the farmer who has raised a second crop of paddy in anticipation of this water. In many cases, by the time the water reaches the last point, by passing from field to field, the supply from the reservoir would have been terminated. Interpersonal rivalries between farmers also aggravate delays. The resultant poor management of irrigation water has been considered one of the important reasons behind the inability of irrigation to stabilise the productivity of paddy (CWRDM 1981).

In the study area while field channels have been laid, they stop short of their full length. As a result, after a point, water is supplied by field-to-field irrigation. From the sample of large, medium and small farmers interviewed, a higher percentage of small and medium farmers are affected by incomplete field channels than the large farmers. They therefore have to rely on field-to-field irrigation. While only 17% of large farmers have to wait for water to flow from field-to-field after the point at which the field channels ends, the percentage is much higher for medium and small farmers, being 41% and 66% respectively. As mentioned in Chapter 4, these percentages have been calculated separately for each landholding group.

TABLE 5.1 Percentage of large, medium and small farmers who have to rely on field-to-field irrigation

Farmer Category	*Percentage in each category affected by incomplete field channels*
Large	17
Medium	41
Small	66

Source: Field Survey 2002

It needs to be noted that by virtue of the larger extent of their holdings, once the water reaches one field, large farmers are able to take it to all the other fields, depending on the water requirement of respective fields. The holdings of small farmers on the other hand are small and dispersed. Take the case of Narayanan, whose entire holding of 0.6 ha is located in three different locations. Two of the plots get water by the field-to-field system, after irrigating the lands of four/five farmers. Hence it is a full five to six days after water begins to be supplied through the field channel, that his fields get irrigated. Devan's plight deserves equal attention in this regard. It takes two nights and one day for the water to irrigate four acres of a large farmer in the area and to fill his pond, the Poricholam kulam, after it reaches Devan's 50-cent (0.2 ha) holding. Devan too has a small share (one-sixteenth) in this pond. When he is too desperate to get the water, he spends money on diesel and pumps water from the pond, as waiting for the canal water may cost him his crop.

It was repeated delays that arose from field by field irrigation that prompted Hamid to purchase an additional 0.2 ha of land located at the point where the field channel ended, despite being burdened by financial problems. When the previous owner decided to sell this particular piece of land, Hamid purchased the same, by taking back the money paid as dowry to his daughter at the time of marriage. Similarly, while Ramachandra Iyer, a large landowner in Cheerani sold his brother's share of the family property following the latter's death, he retained 0.6 ha of his brother's share, as it comprised the plot of land where the field channel ends. The fields that lay further below had to resort to field-to-field irrigation.

Efforts to extend the field channels

In many parts of the ayacut, farmers have striven to extend the field channels to their entire length. Three such cases were observed in the study area, where farmers had been repeatedly requesting the offices of the irrigation department and the CADA, but no action had been taken. In all the cases, the farmers whose fields were located immediately below the present end point of the field channel were resisting the extension of the field channels. At stake in all these cases were ponds that had to be filled up with canal water. Ramachandran (see Box 5.1), whose fields are located far below the ending of the field channel, had to wage a battle with the owners of the Velampotta kulam during every second crop season. The field channel presently ends above the pond, and the flow in the field channel is weak until the pond has been filled. Ramachandran and his neighbouring farmers who cultivate a total of 8 ha have been requesting the CADA dept to extend the field channel up to their fields. The CADA cites inadequacy of funds as the reason for not taking up this work. As a result, Ramachandran is forced to send his workers at night and to block the flow of water into the Velampotta kulam. Similar is the case of a group of farmers in Cheerani whose lands are located 500 metres below the present terminal point of the field channel. Their request for extension of the field channel for an additional 500 metres has been opposed by a large farmer whose lands and ponds benefit from the present arrangement. The large farmer in this case is unwilling to surrender land necessary for the extension of the field channel. Owing to their desperation, the deprived farmers (six of them cultivating a total of 6 ha in the area) offered to pay the price quoted by the large farmer in compensation for the land to be surrendered. The large farmer has however not relented. In the third case, the secretary of the Pulikalpotta CADA committee himself has been pressing for the extension of a field channel. The extension had been sanctioned on two occasions by the CADA department only to be shelved later on. On both the occasions, it was resisted by a large farmer who was unwilling to surrender the land for the necessary extension. On the second occasion, the work was stalled after the sand for construction had been dumped at the proposed site. After pressing for this extension for ten years, the secretary of the CADA committee dug a well on his land bordering a small stream so as to provide him some relief during acute water shortage.

The small depend on the large

Infrastructure in disrepair, a mostly indifferent bureaucracy, and a distribution system that is fairly unregulated sum up the situation. The previous sections have shown how in the absence of a rule enforcing irrigation department, the existing systems of water distribution in the ayacut area reflect the prevailing power relations, with the small and marginal, and women farmers playing a marginal role. This is reflected in the fact that while the economically and socially powerful farmers make their presence felt along the main and distributary canals, the economically less powerful farmers confine their activities to the field channels. They are therefore poorly represented in the 'zone above the outlet' (Mollinga 2003:204).

All the small farmers, and about half of the medium farmers covered in the sample confine themselves to guarding the outlet that supplies water to the field channel that concerns them. Hamid, a farmer with a holding of 1.4 hectares in Velanganpadam says that farmers like him do not go at night to close the shutters along the main canal. They wait for the water to reach their field channel, and may perhaps negotiate with farmers who rely on the same channel, requesting them by saying 'please take water in the other direction later, my crop is drying'. Rajan a small farmer from Kizakemuri, a tail end area, also confines his operations to the outlet that supplies water to his area. 'When farmers from further down come and close the outlet, I hide and wait until they leave, and then open it again.

When they return I hide again'. Farmers like Hamid or Rajan who wield very little social, economic or political clout, may benefit if a large farmer owns land in close proximity. However, they are not in a position to complain when the large farmer first fulfils his demands (which includes filling his pond) before giving water to them, for it is he who has brought water up to their outlet.

Reduced participation of small farmers is also manifest in the activity of getting water into the Chulliar reservoir. As has been detailed in Chapter 3, a series of water transfers from one reservoir to another finally lead into the Chulliar reservoir. While the system is so designed (through the construction of surplus escapes and link canals) that surplus waters from the adjoining Chittoorpuzha basin and the Meenkara dam are diverted into the Chulliar reservoir (see Fig 3.1 in Chapter 3), farmers from the Chulliar command area have to exercise pressure on officials at each point in order to ensure that the Chulliar reservoir is filled. In order to do so farmers from the area hire jeeps, mobilise farmers and go to the Moolatara regulator and the Meenkara dam which are critical diversion points in the interbasin transfer of water to the Chulliar reservoir. While most farmers make contributions in cash towards the expenses required for such an activity, large farmers make larger contributions. They are also found to take a leadership in mobilising farmers to visit various diversion points and to ensure that water is transferred to the Chulliar reservoir. Given the skewed power relations that exist between the large and the small farmers, one cannot expect small farmers like Velan or Paraman to take the initiative of mobilising farmers for this task. The higher presence of large farmers in this exercise is found to give them a moral superiority over the other farmers, which is used to justify their over appropriation of water. Large farmers have voiced this feeling in saying that though they have to resort to under the table measures while opening and closing shutters out of turn, their effort to fill up the Chulliar reservoir entitles them to do so. Many a medium and small farmer also acknowledge the fact that without the effort of the large farmers, water would not have reached the Chulliar reservoir.

If at all small farmers are involved in manipulations along the main canal, it is by being a part of the labour gangs employed by the large farmers. While it is true that large farmers seek the help of small farmers and agricultural workers for the manual work required in closing and opening outlets, the inequality in the relationship is brought out by the fact that while the large farmers provide the money, the small farmers put in the hard work of running around at night. At the end of the exercise, most of the large farmers would have filled their ponds too. Devan, owner of 0.2 ha for instance, helps Mohammed who owns 8 hectares in the same area, in closing outlets and bringing water to the field channel at Ennayaram. Devan stands to gain as his small holding is irrigated by the same field channel.

The dependency of the small farmers on the large farmers is also brought out through the fact that when small farmers face water scarcity, they request the large farmers to pressurise the irrigation department to release water, rather than make an appeal by themselves. The reason they cite is that the department would consider an emergency release of water only if a minimum of fifty farmers needed it. The department would not consider releasing water to save less than a hectare of paddy land. In some cases, the small farmers were found to appeal to the secretary of the padashekhara or the CADA committee who in most cases was an influential person. Such practices reinforce the upper hand that large farmers enjoy in this regard. This is also an indication of the fact that large irrigation systems are not able to cater to the needs of individual farmers, and in such cases, the small farmers bear the brunt.

Coping with an Uncertain Supply of Water

Apart from placing farmers at the tail ends of distributary canals and field channels at risk, the inefficient distribution system also makes farmers undergo considerable uncertainty throughout the second crop cycle. Farmers who do not have access to any privately owned source of irrigation face most of the uncertainty. This uncertainty is found to affect the commencement of the second crop, the method of sowing adopted (broadcasting or transplanting), as well as the timely application of fertilisers.

As has been discussed in Chapter 2, two methods of sowing are adopted during the second crop season-transplanting and broadcasting in slush. The main advantage of the broadcasting method is that it is not labour intensive as seeds can be sown once the fields have been ploughed. This method does not require prior planning as in the case of transplanting wherein a nursery has to be raised in advance and the seedlings have to be transplanted on time. Transplanting is also labour intensive. The critical advantage of the transplanting method is that it enables an early start to the season. This is critical as far as the second crop is concerned for otherwise the harvest extends into the relatively rainless months of January and February (see Table 2.4 in Chapter 2 for the monthly rainfall during this period) when the water demand is the highest. It also allows the farmer to make use of the water that remains in the fields after the harvest of the first crop. Broadcasting on the other hand requires the fields to be drained of water after sowing to enable sprouting, and then irrigated once again by the fifteenth day thereby increasing the requirement for irrigation. In addition, if water is not made available at certain stages of plant growth, the incidence of weed attacks increases. Apart from being a time consuming activity, weeding also increases the costs incurred by farmers towards labour charges. A transplanted crop on the other hand reduces the incidence of weeds. While farmers prefer a transplanted second crop, the uncertainty of water supply makes it difficult to adopt this method. It is now becoming increasingly common for farmers to hurriedly plough and broadcast the seeds once they are assured of water supply from the reservoir.

All the farmers who resorted to transplanting had raised a nursery before being assured of water supply from the reservoir. All these farmers were able to do so as they had access to some alternative source of irrigation. The relative percentage of such farmers in each landholding group is given in table 5.2.

TABLE 5.2 Sowing methods adopted by farmers in each landholding category for the second crop season

Land holding Category	*Transplanting*	*Broadcasting*	*Transplanting and Broadcasting*
Large	52	36	12
Medium	18	77	5
Small	20	80	–

Source: Field Survey 2002

More than half the number of large farmers interviewed had resorted to transplanting, with an additional 12% doing a combination of transplanting and broadcasting. In the case of the medium group, only 18% transplanted, with an additional 5% doing a combination of transplanting and broadcasting. It is significant to note that more than 75% of the farmers in the medium and small farmer category resorted to broadcasting. While 20% of the small farmers resorted to transplanting, this was not due to access to alternative sources of irrigation. 43% of the small farmers who had transplanted did not have access to any alternative source of irrigation. They had raised the nursery in a corner of their fields when it had rained, and had transplanted too during the rains. Another 29% of the small farmers could raise a nursery only because all members in the command area of the pond had raised nurseries and transplanted with the water stored in the pond.

All the farmers in the large and medium farmer category who had transplanted in advance were able to harvest their crop. Large and medium farmers from the tail ends of the second reach in comparison with those in the head reaches however did suffer losses in yield due to inadequate supply of canal water. In the case of the small farmers however 43% of those who had transplanted (the same set of farmers who were without access to alternative sources of irrigation) lost their crop for want of adequate water at later stages of crop growth. Farmers (in all categories) who had not transplanted had resorted to broadcasting in early November after the rains or in late November after getting an assurance of water supply from the reservoir. Amongst this group, those with access to alternative sources of irrigation had sowed soon after the rains in November, while many of those who did not, waited for the assurance of canal water from the irrigation department.

Apart from a delayed start, the farmer is unable to plan the cropping season in advance owing to the uncertainty in canal water supply. In 2001 for instance, some farmers had not even ploughed their fields by October, as there was no assurance of water supply from the reservoir until then. After a few spells of rain, many farmers took the risk of sowing their fields, in the hope that some water would be released from the reservoir in the coming months. On seeing this, many others also hurriedly ploughed their fields and sowed. Farmers observe that once a small section of farmers take the risk and sow, others also follow. This haphazardness is captured in the quote by one of the farmers:

> Those with ponds sow first. Then some others on hearing that there would be some water in the reservoir sow. On seeing this, some others follow, saying: If they can sow, so can we.

This sudden decision to sow on the part of many farmers led to a high demand for tractors, and the tractor owners were more inclined to plough the large holdings first. So many small farmers were not able to plough soon after the rains owing to the non-availability of the tractor.

The uncertainty in water supply also created haphazardness in the method of sowing. During the 2001–02 cropping season, it was late November by when the dam authorities announced that the storage in the Chulliar reservoir would allow for three turns of water supply. Until then many farmers were dilly-dallying on whether to raise a second crop or not. Ramachandran (whose case is mentioned in Box 5.1), for instance had raised a nursery in September. On hearing that there would not be adequate water in the reservoir, he sold the saplings to a neighbouring farmer who had access to a pond. Not having access to even a field channel, he felt that his chances of raising a crop were remote. A few weeks later, on hearing that the reservoir was full, he purchased some saplings and transplanted them on the low-lying lands. But as the cropping season progressed, water got scarce, and he could not apply fertilisers on time. As a result he got a very poor yield. Similar was the case with Velayudhan who owns four hectares of land in Peringotukavu. When he got to know that there was water in the reservoir, he purchased saplings. These saplings however turned out to be over mature for transplanting as a result of which he suffered a poor yield. Another farmer K Ponnan owning 2.4 hectares of land, keeping in mind the debacle of a failed crop the precious year, sowed some of the land with beans. The beans however could not withstand the rains in the month of November, and began to rot. This was the time when it became clear that there was enough water in the reservoir to be supplied to the second reach as well. Ponnan ploughed his lands once more and broadcast the paddy seeds. By then it was early December. This considerable delay coupled with the non-availability of canal water since January gave him a very poor yield. At the end of it all, he neither got a good harvest of beans nor rice.

What emerges is a mismatch between the water demands of the crop and the schedule of water supply. On one hand farmers are unable to plan out their crop schedule. While some of them take a risk and sow or transplant, they may have to put up with unreliable water supplies during critical stages particularly at the time of fertiliser application and at the time of pannicle formation. It is this uncertainty that prompted many farmers in all categories to restrict second crop cultivation to the low-lying fields alone, as potta lands require frequent irrigation.

Table 5.3 Percentage of farmers who cultivated only a part of their holdings during the second crop season of 2001–02

Landholding Category	*Percentage*
Large	31
Medium	27
Small	14

Source: Field Survey 2002

It is interesting to note that compared to 31% of large farmers and 27% of medium farmers only 14% of small farmers restricted cultivation of the second crop. 84% of the small farmers cultivated all of their holdings (2% did not engage in pady cultivation at all). This however does not imply that the majority of the small farmers were assured of an adequate supply of water and hence cultivated their entire holdings. For one, the small size of their holdings makes it difficult for them to cultivate only a part of it. Their disinclination to restrict cultivation can also be attributed to the fact that raising a second crop is important to the family's food security. In the event of not growing enough paddy, they would have to purchase rice from the open market. Hence most of them are seen to take the risk and raise a second crop. The incidence of crop loss is however highest among the small farmer group. This is not the case with many of the medium and most of the large farmers, who consume only a part of the paddy they cultivate, the rest being sold in the open market. Hence, if they are not assured of a fair supply of water, they are less inclined to cultivate paddy on their entire holdings. This trend therefore indicates that the canals are failing in their ability to provide assured irrigation to single cropped paddy land for the second crop season, thereby defeating the very purpose of canal irrigation. Even farmers with alternative sources of irrigation in the tail ends of the second reach hesitate to cultivate their potta lands with paddy during the second crop season.

Institutional Solutions

The two existing institutional solutions aimed at addressing problems with water allocation and distribution in the canal system are the Project Advisory Committee (PAC) and the Beneficiary Farmer Associations (BFAs) alternatively known as Karshaka Samities (meaning farmer associations, 'karshaka' meaning farmer) constituted under thc Command Area Development (CAD) programme. The CADA Act was passed in 1986 in the state, and it envisaged a three tier system consisting of the Project Advisory Committee at the project level, the Canal Committee at the branch/distributary level and the BFA at the outlet level. In this section I examine how the PAC and the BFAs have addressed the issue of unequal water distribution in the command area of the Chulliar system.

The Project Advisory Committee (PAC)

The PAC is the apex body of the three-tier system constituted for every irrigation project. The ex-officio members of the PAC are the District Collector (as Chairperson), the Executive Engineer of the Irrigation Department as Convenor, MPs (Members of Parliament) and MLAs (Members of the Legislative Assembly) whose area of jurisdiction includes the ayacut area of the irrigation project, one official each of the departments of Agriculture and Cooperation, one representative each of the Canal Committees and up to five other members consisting of knowledgeable persons in agriculture and irrigation . The major functions of the PAC include ensuring equitable distribution of water to different parts of the command area in accordance with the water requirement of the crops (Joseph 2001). PAC meetings are usually convened just before the commencement of water delivery during every second crop period (in the months of October/November). At such meetings, decisions are taken regarding the schedule of water delivery to be followed in the coming second crop, keeping in mind the storage in the Chulliar reservoir.

While the PAC was constituted to ensure greater equity in the distribution of water between the first and the second reach of the command area, farmers from the second reach do not feel so. They allege that before the commencement of the second crop season, the PAC routinely advises the second reach farmers not to raise a crop as they stand the risk of facing crop failure. They complain that the PAC does not give any direction to the Irrigation Department to correct the anomalies in the distribution system such that the second reach receives adequate water. On occasions when the farmers from the second reach have appealed to the PAC members to intervene, they allege that their appeals have not been taken note of.[7] They also allege that powerful farmers from the first reach ensure that the PAC does not take any decision that upsets the current status quo. In addition, the manner of constitution of the PAC (the presence of the Collector and other Irrigation officials) makes it quite inflexible in functioning. If an emergency meeting has to be convened to supply water to the second reach, it can only be done depending on the convenience of these officials. The PAC therefore has reduced its functioning to the holding of mandatory meetings prior to the commencement of the second crop season, making no attempt to intervene in the issue of water distribution while the season is in progress.

During the second crop season of 2000–01 and 2001–02, there was never an instance when the irrigation department or the PAC sought to intervene in the distribution of water such that water reached the deprived farmers. The only instance when some corrective action was taken was in December 2001 when a shutter was slashed open by a first reach farmer when water was being supplied to the second reach (see p. 201 of this chapter). The District Collector was sympathetic to the concerns of the second reach farmers. In addition the MLA also pressed for the same. As a result, punitive action was taken against the defaulter, and water supply was stopped by five in the evening, to be resumed only after the shutter was back in place. It needs to be noted that it was not the irrigation department that initiated the action, though it was very well within the powers of the Executive Engineer of the irrigation department to do so. It was only because of the pressure exerted by the District Collector and the MLA concerned that action was taken. When water was supplied to the second reach in the following month, a maestry was posted at Virutti in the second reach, as per the special orders of the Collector to supervise that water was being supplied to the second reach as per schedule. Similarly, deprived farmers were found to appeal to the District Collector and not to the engineers of the Irrigation Department, as they were more hopeful of action being taken by the former than the latter.

The karshaka samities

The karshaka samities or the BFAs were constituted to aid the implementation of the CAD programme, which was intended to facilitate efficient and equitable distribution of water below the outlet. Its members include farmers who own land in the command area of the concerned field channel. These members elect an executive body of not more than seven members from amongst them, consisting of an elected President, Secretary and other office bearers.

The CAD programme grew out of the finding that underutilisation of canal water was primarily caused by inadequate land preparation for irrigation through levelling and shaping of fields as well a lack of field channels and drains, that hampered the flow of water to all fields (Hart 1978). The CAD programme that was implemented in the state therefore laid out a list of activities to be undertaken in the command area of each field channel. This included land levelling, land shaping, development of groundwater for conjunctive use, introduction of suitable cropping patterns and preparation of individual farm plans for farmers (GOK 1981b). The above activities comprise the non-recurring functions of the BFAs, i.e. those which need to be discharged only once. The recurring functions of the BFA include conflict resolution among farmer members, facilitating an equitable distribution of water and sensitising farmers about the need for adoption of group farming (Joseph 2001). This includes proper upkeep of the field channels and other infrastructure, collective procurement and distribution of seed, fertiliser and simple agricultural implements, organising group nurseries and group cultivation, fixing an appropriate time for commencing cultivation, pesticide application and fixing an appropriate crop-pattern for the area (ibid.).

Most of the BFAs in the ayacut area were constituted during the 1988–89 period. The president and secretary were mostly farmers from the large and medium category. The small farmers were largely unaware of its functioning, except when fertilisers were supplied through the BFAs at subsidised rates. Some of the farmers recall that the irrigation officials pushed through the formation of these associations in a hurry, in order to have them in place before a certain date. Critiques of the CADA exercise in the state also point to this issue, wherein the constitution of the BFAs were hurried in order to meet the time schedule of international lending agencies and to avoid funds from lapsing (Joseph 2001). Hence, while the laid out mandate of the

BFAs aimed to set up a participatory style of functioning involving collective action amongst the constituent members, in practice, they exhibited very low levels of participation. The members were also unaware of the broad span of activities that could be taken up by the BFAs. While the CAD programme had envisaged the preparation of individual farm plans and redesigning of cropping patterns, none of the existing office bearers were briefed about such objectives. This point to a significant lack of communication between policy makers, government officials and the farmers, and to significant lapses in the translation of policy into practice.

In practice the entire effort was confined to the development of infrastructural works at the terminal level, viz. the laying of field channels. Even these activities were not undertaken by the BFA directly, but through private contractors. This again has not been completed (as discussed earlier in this chapter) despite the Chulliar irrigation system being functional for more than three decades. Even the regular maintenance of field channels that was supposed to be undertaken by the BFA members, was in practice undertaken by contractors.

Finally, the CAD effort, as in other parts of the country, was focused on the technical and operational features of water distribution, sidelining the institutional aspects of the same (Jairath 1985, Pant 1981). Laying of field channels and organising group cultivation was considered to ensure equitable distribution of water. Internal power relations amongst members of the same association, which play an important role in the final distribution of water were ignored. As cited earlier in this chapter, certain field channels are permanently breached in order to fill ponds located at the head end of the field channel. Such a system of distribution seems to have stabilised over time. In all such cases, the ponds belong to large landowners. Such malpractices are not addressed at the level of the BFA. The fact that field channels have not been extended despite repeated requests indicates the presence of power plays in the distribution of water through field channels and the ineffectiveness of institutions such as the BFA.

Conclusion

The attempt in this chapter has been to illustrate the inequities in water distribution in the Chulliar canal system and the factors that contribute to it. A canal system that has been functional for a mere three decades is currently in a state of disrepair. The problems that arise from the poor state of art of the regulatory devices and the partially constructed field channels cannot be addressed through repairs and maintenance alone. The infrastructure in disrepair is also an indication of power plays along the canals, both above and below the outlet. This is manifest in shutters being removed at will, and locks and keys in the custody of farmers and not *maestries*. It also appears that the missing shutters come to the aid of farmers who wish to take water out of turn. It is easier to plug an open outlet with sand bags and hay rather than raise a locked shutter.

Towards the end of my fieldwork in 2002, I learnt of a survey that had been undertaken by the Irrigation Department to assess the state of art of the outlets and shutters in the canal system. Many farmers were present during the survey, and some of the farmers whom I spoke to were hopeful that the department might initiate some corrective action. An important question that emerges is that if the regulatory devices were to be repaired and made functional, how would one ensure that water distribution follows the laid out schedule? The role that the maestry plays is critical in this respect. Over the years, he has been found to acquiesce with the interests of the economically powerful farmers in the area. The maestry's withdrawal from active management of water distribution has been accepted by both the farmers and the irrigation department. During the past decade when farmers in the second reach have been consistently deprived of water during the third turn, the department has not ensured that the maestry discharges his duties. Not even when the farmers in the second reach filed a legal case against the department for neglecting their water requirements. Departmental inertia to correct the distribution system is also reflected in the fact that canals do not get lined and field channels do not get extended despite repeated requests. It is not enough therefore to line the canals, to restore the outlets to their original dimensions or to put back the forcefully removed shutters. Also required is an overhauling of the functioning of the irrigation department.

Across the country, the department's inability to ensure fair and efficient water distribution has strengthened the argument in favour of farmer's participation in irrigation management. What are the prospects for a PIM programme to improve water distribution in the Gayatri Irrigation Project? The state's irrigation strategy for the Tenth Five Year Plan (2002–2007) proposed to introduce Participatory Irrigation Management in selected projects (GOK 2005). Of the two pilot studies initiated, one was situated in the Malampuzha Irrigation Project in Palakkad district. The main objective was to constitute water user associations who would undertake the operation and maintenance (O & M) of branch canals and distributaries. It would therefore be only a matter of time before the process would begin in remaining irrigation projects in the state. Given the existing power relations between large and small farmers and the poor functioning of the existing BFAs, would the constitution of water user associations under a PIM programme ensure fair and efficient distribution of water? Would the constitution of users associations without over hauling the functioning of the irrigation department ensure the participation of farmers, particularly the small and marginal farmers like Velan or Paraman or Madhavi.

Ironically, while the overall thrust is on enhancing farmer participation in water distribution, all farmers in the tail ends of the second reach have opined that the department take over water distribution fully, and that farmers should not be allowed to intervene. They argue that farmer interventions will result in an inequitable distribution, especially as large farmers in the first reach and those at the head ends fill their ponds with canal water. These issues need to be considered before a PIM programme is implemented in the Chulliar command area in the near future.

Despite the inequitable access to ponds, wells and streams, had canal water been made available in equal amounts to all farmers, some part of this inequity could have been redressed. However, the poor state of canal infrastructure, the corruptible irrigation bureaucracy and the operation of power plays has only aggravated the inequities precipitated by the land reforms as far as access to water is concerned.

The large farmers like Kochukuttan or Ajayan who benefited the most from the land reform exercise continue to benefit the most from the supply of water through the canals.

Notes

1 These include the Malampuzha, Gayatripuzha, Chitturpuzha, Walayar, Kanjirapuzha, Pothundy and Mangalam irrigation systems.

2 Water to the Right Bank Canal of the Gayatripuzha Project is supplied from the Meenkara reservoir alone. The command area of the RBC is 2389 hectares.

3 During 1975–77, the government of India declared internal emergency throughout the country. Sankunni Master recalls that people were afraid of breaking rules during this period, fearing that the administration would come down heavily upon them.

4 The sand bags used to plug the outlets are referred to as 'urea sacks' ('*urea chaaku*' in local parlance), as sand is filled in sacks in which urea is supplied.

5 Keeping in mind the conveyance losses and losses due to ineffective distribution system, a new scheme for revamping and consolidation of 10 old generation projects in the state had been launched during 1997–98 (GOK 2003). Out of the ten selected projects, five were located in the Bharathapuzha river basin, which included the Gayatripuzha Irrigation Project.

6 During the second crop season of 2001–02, water was released in three turns in the months of December, January and February. While the second reach was also supplied water during third and last turn of water release, commencing from 2nd February onwards, the supply did not last for the scheduled seven day period.

7 Suresh, son of Chandran, a farmer from the second reach reports of one such incident in 1998. The crops were drying and they were waiting for the water to reach them. They heard that day that the dam would be closed by evening. So Suresh along with some other tail end farmers rushed to Sivaramakrishnan, an influential farmer from the mid reaches of the area, who was also a PAC member. They appealed to him to ensure that the dam remain open for one more day so that the water reach them. They report that Sivaramakrishnan refused to pay heed to their appeal saying that tail end farmers are forever complaining. Suresh angrily recalls that Sivaramakrishnan could afford to say so as his

crops were thriving with water. He commented that Sivaramakrishnan's response was not befitting of a member of a body that was constituted to ensure justice to all parties concerned. To the contrary, he behaved as though the Chulliar reservoir was his private pond.

6

Floating Ownership Claims

The present chapter looks into how the existing fluid property classifications over water compound the inequities discussed in the previous chapters. After a brief review of the nature of public and private rights to water, the chapter looks at the property-technology interface and how it shapes access to water. A critical discussion lies around the increase of private control at the cost of public and common rights, and how this shapes conflicts and spaces for contestation. In conclusion, this chapter explores alternatives to the existing private and public modes of water resource management.

What is Public and what is Private?

The shifting property status of different sources of water has been captured well by Meinzen Dick in the following paragraph:

> Canal water changes from state property in the main delivery system to common property of a group of farmers on a watercourse, to individual property as it moves on to a farmer's fields, and to an open access as it percolates into the aquifer. This complexity is further increased by the growing use of groundwater, captured and pumped by those who can afford. Once lifted, groundwater can be public, private or common property, depending on who owns the wells, though in practice private ownership dominates (Meinzen-Dick 2000).

This captures the movement of water through different property regimes in the study area as well. There is an added dimension however, that of publicly owned canal water moving on from the field channels into privately owned ponds and into commonly owned streams and rivers, which is again lifted out for irrigating riparian lands.

While the existing legal categories indicate the degree of access that farmers exercise over the use of water, they by themselves do not fully explain the inequities in access to water. Benda-Beckmann makes a distinction between categorical rights, which refer to the broad conceptual legal categories, assigned to resources (such as the existing four main property categories) and concretised rights, which refer to the way in which categorical rights are embedded in the immediate socio-economic context. Rights over the means of appropriation (money and technology for instance) and the existing distribution of wealth that determines such rights are manifestations of concretised rights (Benda-Beckmann 2001). In the following sections I show how a combination of categorical and concretised rights contribute to the presently skewed access to water.

Public property in the Varayiri: The canals and the streams

Water supplied from the Chulliar dam through the network of main and distributary canals, brought to fields through field channels is considered as public water, or '*sarkar vellam*' (meaning water that belongs to the government). In terms of access to water it implies that all farmers in the *ayacut* area are entitled to receive a share of the water. Misappropriation of this water can therefore be viewed as a violation, and can be contested by the deprived users. The infrastructure (viz. the reservoir, canals and regulatory devices) is the property of the state, specifically, the property of the state irrigation department.

Another source of irrigation that is classified as public/common is the stream. The stream can be viewed as a common property resource of riparian landowners, for they stake the first claim to the water in the stream. Non-riparian landowners can take water from the stream (either by pumping or by diverting the water) only through the lands of riparian landowners, for which they require the latter's consent. Since riparian landowners do not allow channels to be dug through their fields to convey the water, non-riparian landowners have to irrigate the formers land before irrigating their own. If the water is being pumped out of the stream, the non-riparian owner has to therefore incur the pumping costs of irrigating both the riparian land and his or her own land. The seeping in of canal water into the stream channel has however led some farmers to argue that non riparian farmers are equally entitled to a share of this water, since it is canal water and not the water in the stream. Reasoning of this kind has prompted the formulation of two lift irrigation schemes, which intend to lift the water from the river and store it in private ponds that are not located on riparian land, from where it could be distributed to the lands below.

The stream can also be regarded as public or state property as the use of water from the stream for irrigation purposes is regulated by the state. As has been discussed in Chapter 3, individual pumping of water from the stream for irrigation purposes is forbidden. This restriction is evaded by farmers who dig wells on riparian land, depicting the privately owned well as the source of irrigation and not the stream.

Ponds as both private property and common property

Water in the pond is viewed as the private property of those who own land in its ayacut. They are better classified as shared private property or as common property of the concerned landowners. These right holders have the right to exclude non–members from taking water for irrigation. The catchment through which water flows into the pond is primarily composed of private land. In most cases however those who own land-owning rights in the catchment need not be those who own landowning rights in the command area of the pond. Hence, rainwater when channelled into the pond becomes private property of those who own land in the ayacut area of the pond. One possible reason for the pond being considered private property is the fact that most ponds were constructed by individual landowners in the pre land reform era, the erstwhile janmis, to irrigate their paddy lands. Following the reorganisation of land and water rights through the land reforms of the 1970s, the pond continued to be viewed as private property, despite the fact that none of the present owners had contributed towards the construction of the pond. Similar to ponds, shallow pits (kuzhis) and wells (both shallow and deep) are treated as private property as well. However, unlike ponds, these irrigation sources belong to a single owner in most cases, or to a family.

Private ownership of ponds implies that farmers who own land in the pond *ayacut* have sole rights to the use of water for irrigation. They are not bound to share this water with other farmers; if they do so, it is only on the basis of personal or other affinities. The superior rights of the landowners in the command area of the pond also implies that irrigation takes a priority over the numerous other uses of pond water. These include livelihood needs like fishing, picking of lotus tubers and washing of clothes by washer folk, as well as human needs like washing clothes and bathing. Fisher folk cannot claim rights to fish in private ponds. Fishing is mostly undertaken at the end of the irrigation season and the catch is shared amongst the right holders.

There are only few instances when non-pond owners use water from ponds to meet their livelihoods. One is when the pond is owned by the panchayat and not by an individual or group of individuals. In such cases, those who cultivate land in the *ayacut* do not enjoy a priority over others in taking water for irrigation. More important, people in the vicinity can demand that their needs for bathing and washing clothes be given a priority over irrigation needs. The panchayat acquires ownership over ponds only when private owners surrender the same to the panchayat; no ponds have been constructed by the panchayat. The only instances of pond surrender have taken place during the implementation of the land reforms when large farmers in an effort to evade the ceiling surrendered the land on which ponds were situated.

The other instance when ponds have been used to meet livelihood needs other than agriculture is when the tubers of the lotus growing in certain ponds are picked by tuber pickers. A group of tuber pickers from the nearby village of Pallasana was observed to go from pond to pond in the area, picking lotus tubers in the months between January and June.[1] They belong to the Tamil speaking Chettiar community. Tuber picking was an activity that only members of their community engaged in. These lotus tubers were then dried and made into 'kondattam' (an eatable). Lotuses were found to grow mostly in ponds that were not used for bathing by the local people. This however does not imply that lotus grew in all the ponds that were not used for bathing. The tuber pickers would pay a certain amount to the pond owners for being allowed to pick the tubers, and they would pick the tubers twice during this season. They paid the owner of the Choorikad kulam for instance one thousand rupees for the period between January and June. Each person would be able to pull about forty to fifty kilograms of tubers, which they would sell at about rupees twelve per kg in the market in Palakkad. It is a lot of hard work, for one has to stand in thick slush and feel the tuber with one's feet and then pull it out. They would stop this activity with the onset of the monsoons in June. Then once the plants flowered, they would go and pick the lotus flowers.

Focal points of inequity: Private tubewells and private ponds

In this section, I discuss how skewed access to private irrigation sources namely ponds and tubewells, exacerbate inequities in access to water. I shall take up the case of the tubewell first.

TABLE 6.1 Percentage of farmers in each landholding category with access to tubewells

Land Holding Category	*Percentage with tubewells*
Large	35
Medium	4
Small	–

Source: Field Survey 2002

Private tubewells

While 35% of the sampled large farmers sampled owned tubewells, only 6% of the medium farmers did. None of the small farmers had dug tubewells. Of the 35% of large farmers who did, 18% had dug more than one tubewell. It is also important to note that the medium farmers who had invested in tubewells could do so only because of the availability of a non-farming income, such as earnings from running commercial enterprises or remittances sent by family members working outside the country.

Amongst the large farmers a new trend that has been noted is that of setting up an interlinked system of ponds, tubewells and shallow pits. Ponnunni (owner of ten hectares of land in Matacode) for instance has dug a tubewell to a depth of 300 ft. During the second crop season, his 10hp motor is found to pump water continuously which is then stored in a pond to which he has exclusive water rights. This pond is found to be brimming with water when ponds around display their dry beds. Bhaskaran, a large farmer who owns land in the tail reaches of the Peringotukavu distributary canal, disillusioned by the repeated failure of canal water was in the process of digging two tubewells in 2002. He told us that he was investing a high amount on the tubewells, as he was adamant upon raising two crops of paddy. His plan was to store the water in the three ponds that he owned (of which he had exclusive rights to two). The increased storage in one of these ponds was also expected to raise the water levels in an adjoining shallow pit on his land. He was hopeful that by pumping water from all these water sources, he would be able to avert crop failure during the second crop season. Tubewells are therefore becoming an important 'exit option' (Wood 1999) for farmers faced with an unreliable canal water supply. In fact all the tubewell owners in the large farmer category stated that they are thankful that they do not have to run after canal water like the other farmers. Tubewell owners are also found to be less concerned about the state of art of ponds in which they have water rights. This is an interesting phenomenon, the way in which the emergence of an exclusively private solution like the tubewell reduces the farmers' interest in the upkeep of the ponds, which can be viewed as the common property of all those who own land in the command area. Such a trend has been noted from other parts of south India too wherein the growth of privately owned and operated tubewells is being viewed as a threat to the survival of existing common property resources such as ponds (Gunneli and

Krishnamurthy 2003). In the study area it has been pointed out that unless the ponds are owned exclusively by the farmer concerned, once a farmer digs a tubewell that yields him water, he is not inclined to engage in pond maintenance and in cleaning the channels that direct surface water into ponds. The new tubewell owners were also found to be less inclined to engage in manipulations along the canal that would lead canal water into the pond.

The sustainability implications of these exit options have also not merited any attention. Chapter 3 has discussed reports of tubewells in certain areas lowering the water levels in the shallow wells in the vicinity. In one particular case observed in Velampotta, a new well had to be dug for drinking water supply as the older one had dried up since the digging of tubewells by a large farmer in the vicinity. The sustainability and equity dimensions are intertwined. While large farmers dominate over the ownership of tubewells, shallow open wells are mostly owned by medium and small farmers. The drying of open wells therefore results in reduced access of the medium and small farmers to water. This phenomenon has been widely reported from different parts of the country (Reddy 2002). The extent to which the extraction of water from privately owned tubewells reduces the water levels in shallow wells and ponds or reduces the base flows in streams (all of which are not exclusively private sources) is an issue that has important implications on equitable distribution of water. The inequity is more sharply felt when affluent tubewell owners with high pumping capacity resort to water sales. Groundwater sale was observed in the earlier mentioned Ponnunni's case. In this case, the groundwater was pumped and stored in a pond, from where it was sold to farmers who wished to raise a third crop of vegetables in the months of March–May, the driest months of the year. Ponnunni sold the water to Aru, an agricultural labourer who worked for Ponnunni. He made it appear to Aru that he was paying for the electricity charges incurred while pumping and not for the water per se. While only one such instance of groundwater sale has been observed so far, it is an indication of similar instances of water sales in future.

Private ponds

As has been discussed in Chapter 4, access to ponds is not uniformly distributed amongst farmers. While 89% of large farmers enjoy access to ponds, the figure comes down to 48% and 23% in the case of medium and small farmers. In addition, 37% of large farmers have access to more than one pond, while the corresponding figures are only 8% and 6% for the medium and small farmers (see Table 4.2 and 4.3 of Chapter 4). It needs to be noted however, that along with shared ponds, there are a number of ponds that are exclusively owned by a single individual or as is more common, by a single large family. As mentioned in Chapter 4, owners of such ponds enjoy greater flexibility while taking water for irrigation, and the chances of dispute amongst pond owners is reduced. While 67% of the large farmers have access to such family owned ponds, only 10% of the medium and a mere 2% of the small farmer group benefit from the same.

This inequity is aggravated by the fact that these ponds are filled with canal water at least twice during the second crop season. The deprivation of non-pond owners is manifest in two ways. One, they are deprived of the canal water that gets stored in private ponds. Two, the time taken to fill ponds prolongs the waiting period for those downstream. Table 6.3 of this chapter indicates that almost half (43%) of the small farmers suffered from poor crop yields. This was largely due to non-availability of water at critical periods of crop growth. Had this section of farmers enjoyed water rights to ponds, the percentage suffering from poor yields would have come down. The corresponding figures for poor crop yields suffered by large and medium farmers were 0 and 4% respectively, indicating the disparity between those who have access to ponds and those who don't. The following statements made by farmers indicate the critical role played by ponds:

> 'Without ponds everything comes to a standstill' (Kochukuttan, a large farmer).

> 'Those who own ponds can sleep well at night. They have something to bank upon' (Narayanan, a small farmer).

Those without access to ponds argue that ponds should be declared as public property for the following reasons. They argue that by being privately owned receptacles, ponds buffer the farmers in the command area of the pond against crop failure, leaving the non-pond owners to face the brunt of water scarcity. They also argue that by hoarding public water, ponds exacerbate scarcity for the non-pond owners.

The Property Technology Interface

Most of the conceptual and legal categories that we now deal with (public, common, private and state owned) have been handed down from an era when the extractive power of technology was limited. The extent of access granted by these traditional classifications has however been radically transformed with the advent of modern drilling techniques and energised lifting devices. Since the latter has not been subject to any property categorisations and restrictions, they have helped to strengthen private control over resources. In the study area, while these modern technologies lead to limitless exploitation of water from privately owned water sources such as tubewells and ponds (particularly when the latter are owned by a single individual or family), they are capable of transforming public water sources such as streams into open access ones.

In the case of the tubewells, the availability of energised bores along with the prevailing perception of groundwater as purely private property of the concerned land owner creates a situation of uncontrolled exploitation of water. This is further aggravated by the absence of any institutional check on the volume of water extracted (Hardiman 2007). In the case of groundwater it has been argued that while the existing laws tied rights over groundwater with ownership over land, it did not tie the amount of land owned to rights over volumes of water (Dubash 2007, Bhatia 1992). To the contrary, with the use of land as collateral, state sponsored credit arrangements introduced a de facto relationship between the amount of land owned and the volume of water controlled (Dubash 2007). The combination of modern technology and private property over water is best illustrated in the earlier mentioned Ponnunni's and

Bhaskaran's case, who by storing groundwater in ponds create private reserves of water.

In the case of public resources such as streams, the operation of energised lifting devices in a climate of unregulated use gives those with increased 'pumping power' an upper hand. The wealthy farmers install electric pump sets along the riverbed, and pump water till the stream is dry. Arumukhan, a large farmer in the area for instance has set up an elaborate pumping mechanism from the stream and the river. He has an electric pump set installed along the Gayatri River, from where the pumped water is brought through a pipe into the Varayiri stream bed, which is a tributary to the Gayatri. At this point, the water is once again lifted out through yet another electric pump set and stored in his pond, from which water is taken to his fields. Arumukhan's pond at Tekkinkadu therefore is always full of water. Small farmers like Chellan who own a mere 0.2 hectares of land, construct mud bunds across the stream to divert water. At times these mud bunds are demolished by downstream farmers at night. At other times, the flow in the stream is too weak, and then he has to resort to pumping, by hiring an oil engine. Inequities therefore largely emerge from the differential ability to invest in pumping.

In the case of ponds, the dynamics are different. There appears to be some kind of informal understanding that all landowners have a stake in the water which prevents one farmer from pumping too much of the water. At the same time, farmers from certain pond *ayacuts* have reported that large landowners have pumped excessively. This is particularly the case when the large farmer concerned owns the pond as well. Once again large farmers are less restrained by the financial costs incurred in the pumping exercise than are small and marginal farmers.

The haves and have-nots

Pumping being expensive, inequities in water use arise from the differing purchasing power of the well to do and less well off farmers. Since not all farmers can be 'self providers' with their own pump sets (Wood 1999: 782), inequities result. Wilson for instance has analysed the distribution of diesel pump sets across landholding size in Bihar to capture inequality on this front (Wilson 2002). Even amongst the so-called self-providers in the study area, owners of electric pump sets are at an advantage when compared to owners of the oil engines. The daily operating costs of electric pump sets are lower than oil engines as electricity was supplied free to paddy farmers until 1999–2000, irrespective of the size of the land holding. This norm was later changed, according to which only small and marginal farmers (defined as those who owned less than 2.5 hectares of paddy land) were exempted from paying tariff. Those who owned dry land below two hectares were also exempted. When the present study was being conducted, confusion prevailed over the exact norms and rules in this regard.

The initial installation costs are however much higher for the electric pump sets, which includes the purchase of the electric pump set, the construction of a motor house to house the pump set and electric wiring. In some cases, additional electric posts have to be installed if the site of pumps is removed from the existing line of electric supply. In addition, its installation requires bureaucratic sanction from the electricity and agriculture departments, with it being reported that bribing the officers concerned is often a must for a speedy installation (see p. 128 in Chapter 3 for the procedure followed).[2] This is particularly so when the water source in question is not a private water source. As per the existing irrigation laws, pumping of water from streams and rivers is forbidden. This restriction is evaded by digging wells on riparian land and showing the well as a private water source. As a result, it is only the large and well to do farmers who are able to afford electric pump sets. Most medium farmers own oil engines, which are fuelled by diesel or kerosene. Most of the small farmers who own oil engines have been able to do so only because of the subsidy made available to them from the government. Many medium and all small farmers are found to hire oil engines rather than own one.

The costs incurred by farmers who have to hire oil engines ranged between rupees fifty to rupees seventy per hour of pumping, which included the rental charges for the pump sets and the cost of the fuel. Most medium farmers and all small farmers who had to hire oil engines were found to calculate whether this amount was worth spending. As a result, all small farmers were found to invest money in pumping only when it was certain that without it the crop would fail. Most often pumping was resorted to at the flowering stage when water availability was critical. Not a single small farmer pumped water in order to apply fertilisers on time. While they knew that not doing so would reduce their yields, they felt that the costs of pumping did not justify the benefits.

The case of small farmers like Paraman and Velan make this clear. Paraman, who works as the supervisor of the land owned by an absentee landlord in the area cultivated his 23 cents (0.092 ha) of paddy land during both seasons. His field is located in close proximity to the Talachera kulam, but has no water rights from the pond, as the land was given to him as excess land at the time of land reforms (see Chapter 4 for a discussion on how those who were given excess land were not given water rights). In order to pump water, he has to pay Rs 100 per hour to the owner of the pond. When the crop reached the panicle formation stage, the canal water had not reached his fields. Were he to pump he would have to do so for at least eight to ten hours, and at that point of time, he did not have the money with him. He finally lost his crop.

The case of Velan also illustrates the travails that a small farmer has to go through to ensure a crop of paddy. The 60 cents (0.24 ha) of land that Velan's wife inherited is located close to the Kuttikadu kulam. The owner of the pond (for whom the earlier mentioned Paraman worked as supervisor) had not raised a second crop and hence there was some water in the pond. Velan along with another small farmer Krishnan, and a medium farmer Chentamarakshan, who owned about 0.8 hectares of paddy land, requested Paraman to allow them to pump water from the Kuttikadu kulam. The trio had hired an oil engine and had purchased some kerosene. Velan's land was located the furthest away from the pond, and so the fields of Krishnan and Chentamara had to be irrigated before the water reached his. Since the water level in the pond has receded, they had to dig a pit in the pond bed before pumping the water. Sixty-year-old Velan was made to do all the manual work, and Chentamarakshan was supervising it. Velan had brought with him one bottle of kerosene, and Chentamarakshan was making fun of him for bringing so little. When I met Velan a few days later at his home, he was telling me that that bottle of kerosene was all that was left of the cooking fuel at home. Since he was unable to share the cost of the kerosene, Chentamara made him do the manual work. After all this effort, Velan got three sacks of grain (about 170 kg of paddy) from his small plot of land.

Very often small farmers are drawn into a web of dependency in their struggle to mobilise the money required for pumping water at critical stages of plant growth. Take the case of fifty cent (0.2 ha) holder Devan in Manalipadam (see p. 209 in Chapter 5). Devan engages himself in fishing and in the 'chittie business', as farming alone gives him little. Devan has water rights in the Poricholam kulam. Devan works as a helping hand to Mohammed, a large farmer who also owns land in the command area of the same pond. Mohammed deputes Devan to purchase pesticides and fertilizers from the market. Devan also runs around with the agricultural labourers mobilised by Mohammed to open and close shutters when canal water is distributed. He also stands guard at outlets that supply water to their area. When Devan is compelled to pump water from the Poricholam kulam, he has to first irrigate Mohammed's lands, as the latter's lands lie before his. While it takes him four hours to pump and irrigate his fields, it takes an additional twelve hours to irrigate Mohammed's fields. While Mohammed is not obliged to share the pumping costs, he allows Devan to use his oil engine for free at times. Sometimes, he may give him some diesel too, not always though. Similarly, while raising a nursery, Mohammed raises a large nursery, and tells Devan that he need not raise one separately for his small field, but can take saplings from the formers nursery. During transplanting, if Mohammed pumps water, he allows Devan to take some water too. Similarly while applying pesticides on Mohammed's fields, Devan sprays the leftovers on his fields. Devan says that by working with Mohammed in such a manner, Devan is able to get water at critical stages. While it does make him dependent on Mohammed, he feels this is the only way in which a small farmer like him can raise a crop without incurring too many costs.

The resultant inequities are reflected in the fact that during the 2001–02 second crop season for instance, while 94% of the large farmers resorted to pumping from streams, ponds and wells to meet their water requirements, only 59% of the medium farmers and 28% of the small farmers could do so. Once again, percentages have been calculated separately for each land holding group.

TABLE 6.2 Percentage of farmers who resorted to pumping of water

Land Holding Category	*Percentage who resorted to pumping*
Large	94
Medium	59
Small	28

Source: Field Survey 2002

Water requirements of paddy show two peaks, one at the time of transplantation, and two, at the time of young panicle formation followed by booting (Narayana et al. 1982). It was during the second stage that most farmers with access to water were found to resort to pumping. Those who were not able to provide water to their crops during the stage of panicle formation were found to suffer from poor yields or even crop loss. Water was also required while fertilisers had to be applied. Hence those without timely access to water were found to apply fertilisers twice instead of the prescribed three times. Thus, while only 4% of the large farmers and the medium farmers suffered crop loss in the second crop season(i.e. they lost their entire crop), 28% of the small farmers suffered crop loss. It needs to be noted that the large and medium farmers who did suffer crop loss were concentrated in the tail ends of the second reach, while the small farmers were more spread out over the region. Similarly while no large farmer suffered from excessively poor yields, 4% of the medium and 43% of the small farmers did. Poor yields in this case has been categorised as less than 550 kg per acre (roughly 1300 kg per hectare). 550 kilograms comprise one truckload of paddy and farmers were found to refer to one truckload of paddy per acre (0.4 hectare) as the lowest minimum yield. Hence, yields were found to range between less than 1000 kg per hectare to 4000 kg per hectare, depending on the degree of access to water at critical stages of the crop cycle.

TABLE 6.3 Percentage of farmers who suffered from crop losses

Landholding Category	*Percentage that suffered from crop loss*	*Percentage that suffered from poor yields*
Large	4	–
Medium	4	4
Small	28	43

Source: Field Survey 2002

Energised lifting of water therefore has not only revolutionised the quantum of surface and sub surface water made available to farmers, it has also widened the gap between the 'resource-rich' (Raju et al 2004: 270, Reddy 2002) and the resource poor farmers in terms of access to water. The dynamics of the property-technology interface also illustrates importance of understanding the manner in which categorical rights are embedded in socio-economic power relations. A combination of both these categories of rights helps to explain skewed nature of access to water.

Contested Water Rights

Farmers have reported quite a number of conflicts over water rights in the area. Some of them were unfolding during the period of fieldwork, but most of them took place earlier. All these conflicts have been focused around the property status of the water source in question. Contestations mostly arise when water moves from one irrigation system to another, and thereby from one property category to another. In this section, I discuss the nature of the contestation in different conflict situations and the measures that people resort to while in conflict.

Asserting rights to a public resource

Most of the conflicts have emerged when a public resource has been appropriated by individual farmers. This usually takes place when canal water is diverted into ponds, when water lifted from streams is stored in ponds, when water stored in panchayat ponds are used for private irrigation and when tail end farmers do not get their due share of canal water and so on. Each of these shall be discussed below.

Canal water in ponds

As mentioned in Chapter 3 and 5, the filling of ponds with canal water is widespread. All farmers admit that it is incorrect to transfer water from the canals into ponds, but all of them engage in this practice. Some of the farmers are more open about it, admitting that 'We know it is not correct, but unless we periodically fill our ponds with canal water, we will not be assured of a good harvest'. Some others are hesitant to talk about it openly. Those without access to ponds have bitterly complained against this practice, for it deprives them of timely supplies of water. This is particularly the case with farmers located in the tail ends of the second reach, where even those with ponds are not always assured of a good harvest.

The irrigation department has however not come out with a clear stand on this issue. Not even when the canal water so stored in ponds has been sold by the pond owners. In all, water sales have been reported from four ponds in the watershed, of which three are located immediately below the main or distributary canals. The underlying issue is that public water that all farmers in the command area of the Chulliar are entitled to, is diverted into private ponds and then sold to farmers whose deprivation is an outcome of this hoarding of water. Such practices have been contested only when the deprived farmer or farmers have protested. While a few such contestations have been reported, there has been no significant individual or public action against the impropriety of the act. Kochukuttan recalls that during the 1980s, one of the farmers downstream of his pond had complained to the engineer of the irrigation department that Kochukuttan's father was diverting canal water into the Pandalamkulam. On finding it to be true, the engineer instructed his father to put an end to the practice. His father however resorted to diverting canal water in the following season as well. This time the complainant filed a case at the police station, charging Kochukuttan's father of misappropriation of canal water and thereby depriving downstream farmers. Kochukuttan went to the police station and bribed the officer concerned as a result of which the case was dropped. Since that incident, he has continued with the practice of diverting canal water, but no complaint has been lodged.

Farmers opine that filing a complaint at the local police station is more effective than filing a complaint in the irrigation department, as people fear the police more than the irrigation department officials. This is also because the department officials so far have not taken any punitive action against farmers for diverting canal water into ponds. The irrigation department officials on the other hand observe that even if they were to reprimand the farmers, they would not be able to stop them from diverting canal water. The impression I get while talking to irrigation officials in this regard is that they do not want to involve themselves in this messy issue. They feel their job is to distribute the water through the canal network and nothing more. They continue to be apathetic even when the irrigation infrastructure has been tampered with in order to facilitate such diversions. It has been reported that in 1995, owners of the Tayamkulam situated by the main canal openly breached the main canal and filled the pond with water which they subsequently sold. Farmers had complained to the police, but it is reported that the owners of the Tayankulam hushed up the case with money. Even at this point the irrigation department did not consider it necessary to take action against the farmers concerned.

While irrigation officials appear disinclined to intervene in such situations, farmers have reportedly bribed irrigation officials in order to take water out of turn and to store it in their ponds. The Mele Cheerani kulam, located in the second reach of the Chulliar *ayacut* for instance (see p. 207 in Chapter 5) was filled with canal water at a time when water was not being supplied to the second reach.

This is reported to have been made possible by bribing the officers concerned. The above-mentioned Tayankulam was also filled in such a manner. Similarly Neelan, a small farmer in the low lying Velampotta area had to resort to siphoning water from the canal into the stream that flowed past his land. He admits of having intimated the maestry and the assistant engineer of his plan to do so and of having bribed them as well. Maniyan, another small farmer has water rights in the Kallankulam, which is located by the side of the left bank main canal at Mannathupara. Since there is no outlet along the main canal in this stretch, he is not able to divert canal water into this pond. Maniyan feels that if he pays the engineer about five thousand rupees, he may agree to provide an outlet at a convenient spot.

While the act of storing canal water in ponds is still considered to be illegal (by farmers and irrigation department officials), its widespread practice has rendered it greater legitimacy over the years. This is reflected in an incident that took place more recently, in the year 2000. Velu a small farmer owns land in the command of the Puthenkulam. Another landowner was Krishnan, an absentee landowner, whose brother in law Ravi managed the agriculture on his behalf. Prior to the commencement of water supply through the canals, Ravi had informed the other right holders that they would have to contribute money towards paying labourers to divert canal water into the pond at night. However, one night, before informing the other right holders, Ravi mobilised a few workers and diverted water into the pond. Velu was away working at the brick kilns in the neighbouring district of Thrissur when this incident took place. On his return, when he attempted to take water from the pond, Ravi objected saying he had to pay for the water before doing so. He demanded that he pay him one thousand rupees, which was the cost of mobilising canal water into the pond. Velu filed a complaint in the police station and on hearing both sides of the story the police officer asked Velu to pay rupees two hundred and fifty. Since he was late in paying this amount Ravi used up most of the water and there was little remaining in the pond. The point to note here is that even the police officer in this case seemed to think that Velu ought to pay Ravi for the act of diverting canal water, thereby legitimising the very act itself.

On the whole, despite the inequity precipitated by such acts, there is general acceptance that this practice is here to stay. This perhaps explains the absence of any major conflict around this issue. The fact that farmers deprived of rights to ponds are dispersed throughout the region must also be preventing any kind of collective action in this regard. Moreover, the small and marginal farmers who enjoy the least access to ponds (23%) are caught in the struggle to meet two ends meet. Hence finding any additional time, money and energy to engage in confrontations is difficult as far as they are concerned. In addition, confronting the large farmers who explicitly engage in the task of diverting canal water is also problematic, as the small farmers depend on the large in very many ways in order to raise a crop of paddy. The other pertinent issue that emerges is that whenever there have been contestations, it is the police who are mediating more than the irrigation department or the panchayat for that matter. The police who are largely unaware of the intricacies of land and water rights therefore take a decision in such matters.

Canal water in the streams

Running water in streams and rivers has always been subject to contestations and conflicts. Traditionally the water in running streams was made use of for irrigation by constructing mud bunds across the stream channel, which enabled riparian farmers to divert the water into their fields. These bunds were a point of contestation between upstream and downstream farmers, particularly when water was in high demand. It was common therefore for bunds to be slashed by downstream farmers. This is also the case with the more recently constructed check dams cum regulators that have been constructed by the panchayat or the minor irrigation department. In at least five cases, downstream farmers had permanently removed the shutters of the regulators. Lift irrigation schemes are also being heavily contested, particularly as they are located along the same stream or river. This has been illustrated in the case of the schemes across the Tekkepuzha, a stream in the adjoining watershed. The implementation of decentralisation has witnessed a spurt in the number of check dam cum lift irrigation schemes, as it is the one way in which the panchayat is able to demonstrate its ability to redress the water woes of paddy farmers, particularly those who are located at the tail ends.

As the number of such schemes increase and the total flow decreases particularly in the summer months, farmers benefiting from the older schemes are likely to assert their superior rights. It even leads to contestations over the canal water that seeps into the stream. While the water supplied through the canal network of the Chulliar reservoir seeps into the Varayiri stream, the water supplied through the canal network of the adjoining Malampuzha irrigation project (a network that supplies waters stored in the Malampuzha reservoir to various parts of Palakkad district) seeps into the Gayatri River. The Varayiri stream meets the Gayatri River at the western tip of the watershed, and hence the lower reaches of the watershed are not far from the Gayatri River. Two lift irrigation schemes that take benefit of the canal water that seeps into the Gayatri River are the Tootipadam lift irrigation scheme and the Cheramangalam lift irrigation scheme. In addition, two new lift irrigation schemes were being proposed upstream of the existing ones in 2001. These were the Peringotukavu and the Tumbidi lift irrigation schemes. Both these schemes intended to pump water from the Gayatri River. Unlike the existing lift irrigation schemes, the newly proposed schemes intended to store the water lifted from the river into private ponds. This was because the water in the rivers would not be adequate for pumping at all times of the cropping season. They also intended to make use of the existing field channels of the canal system to distribute the water. Both the new schemes were being opposed by the farmers benefiting from the Cheramangalam lift scheme further downstream, who went to court in defence of their case. Their argument was that the flow in the Gayatri River alone was not adequate in meeting their demands, and that their needs were met only when the water supplied through the Malampuzha irrigation system reached the Gayatri River. Since their scheme was an older one, they argued that they had a first right to the Malampuzha waters that seeped into the Gayatri River. The proponents of the Peringotukavu lift irrigation scheme in turn were arguing that if the Cheramangalam farmers were indeed benefiting from the canal waters of the Malampuzha system, then on account of its public nature, the claims of the Cheramangalam farmers' claims of having been the first users of this water would not stand. Since it was public water, they too had an equal right to it, they argued. The other argument of the potential beneficiaries of the Peringotukavu scheme is also worth noting. They argued that it was not only the water distributed through the Malampuzha canal network that seeped into the Gayatri River. The water supplied through the Gayatri canal network also seeped into the Gayatri River, and that since they were entitled to the water from the Gayatri project, they had a right to the water that seeped into the river as well. As of 2002, the deadlock had not been resolved.

While these arguments can be carried on inconclusively, the issue of water availability is totally sidelined. Protests over the proposed lift irrigation schemes across the Gayatri River culminated in court cases, which had not been resolved. The conflicts over schemes across the Tekkepuzha were still being resolved through negotiations at the level of the panchayat. One reason why the Cheramanaglam farmers went to court was because both the proposed upstream schemes intended to store the lifted water in ponds. They posed a larger threat than the ones proposed across the Tekkepuzha. The arguments and counter arguments were however focused on the issue of rights alone. At no point did the court order measurement of river flows through the year to assess the total amount of water available for pumping. Such measurements would have been made possible an assessment of the total inflow of canal water into the stream as well, and whether it justified the increase in the number of lift irrigation schemes.

Asserting rights over panchayat ponds

As mentioned earlier, panchayat owned ponds are not the exclusive private property of those who own land in its *ayacut*. As a result, water from these ponds is not taken for irrigation. Of the three ponds in the watershed that are not used for irrigation, two were surrendered to the panchayat during the land reforms. The land on which the Perinkulam in Mannathupara is located was surrendered by its owner Unniappa Menon as 'excess land', in order to avoid surrendering paddy land. Even after the surrender, farmers in the ayacut continued taking water for irrigation. Since it was vested with the panchayat, farmers from outside the ayacut began to take water for irrigation as well. During the drought of the 1980s, farmers from both within and outside the ayacut are reported to have pumped the pond dry. It has been reported that the water pumped out from the pond was taken via pipes to locations that were kilometres away. Despite this all the farmers lost their crop, and the pond was bone dry. This is reported to have motivated the local people to organise themselves into a 'people's committee' to protest against the drying up of the panchayat pond. Since then the panchayat has taken more interest in its maintenance and the water in the pond is stored for bathing. Since the late 1990s, the panchayat has also started a fish-breeding programme in this pond. During the early 1990s there was an incident in which a large farmer in the area pumped water from the Perinkulam at night. Being closely related to the then MLA, he thought he would be able to overcome the local opposition with political power. The local people however were adamant, and the large farmer concerned had to stop the pumping. While his argument was that being a panchayat pond, anybody could take water, the local people argued that the water in a panchayat pond was reserved for the livelihood needs of the local people and not for meeting the irrigation requirements of a few farmers. Another such instance has been observed with regard to the Odungatuchira, which was also surrendered to the panchayat at the time of land reforms. The land in the command area was subsequently sold and the new owner purchased the land on the assumption that he would be able to take water from the pond. On doing so, the local washer folk protested and they are reported to have taken out a protest rally through the town. Their argument was that taking water for irrigation from this pond would jeopardise the livelihoods of ten washer folk families in the vicinity. These are the two instances when the supremacy of irrigation has been questioned through local action.

No fresh instances of pond surrender have been reported from the area. This is despite the fact that on surrender the panchayat would undertake desilting of the pond bed and protection of the side bunds at its expense. Farmers fear that on surrender the pond would become public property, as a result of which they would lose their exclusive water rights over the pond and would have to share it with others.

Apart from lotus tuber picking, washing of clothes by washer folk and fishing are the other small livelihoods that are dependent on the water in the pond. This indicates that irrigation is not the only use of water stored in the pond. It also implies that though ponds are currently viewed as the property of the landowners concerned, customary rights to the water were exercised by the above-mentioned groups of people. Such livelihoods however may be threatened as ponds begin to be viewed as exclusive private property, especially when a single owner owns the pond. There have been instances where farmers are beginning to use ponds for intensive fish breeding. In the watershed, two private ponds and one panchayat pond has been used for fish breeding purposes. Two large farmers who were engaged in fish breeding were contemplating giving up paddy cultivation altogether, as fishing yielded more benefits than paddy farming. One cannot rule out the possibility of other pond owners also resorting to such a measure in future, especially keeping in mind the relative non-profitability of paddy. Such options can however decrease the space available for other livelihood options pursued by the tuber pickers and the washer folk. In the two cases of contestations outlined above, people have resorted to public protests. In both these cases, these public protests have proved fruitful too. In the case of private ponds however, no such rights can be asserted. The pond owner is not even expected to leave some water in the pond to meet the bathing requirements of the local people.

Head-tail conflicts in the canal system

Chapter 5 has discussed in detail the distributional maladies in the canal system. Farmers in the second reach are at a disadvantage, particularly those located at the tail ends of the main and distributary canals. From amongst the farmers in the tail ends, the small farmers are at a particular disadvantage, unable to mobilize labourers to run around for water. Despite the blatant disparity in water supply to the first and the second reach, little corrective action has been attempted.

The lone reported effort was made by farmers located towards the tail end of the main canal during the 1990s. Some of the large farmers in the area, which included a lawyer from one of the prominent land owning families, filed a court case against the irrigation department during the early 1990s for not meeting the needs of the tail end farmers. While farmers both big and small were party to the case, the initiative was mostly taken by the large farmers who also mobilised the financial resources required. As a result, most people attribute the effort put in towards fighting the case in court to the large farmers. The continuing indifference of the irrigation department despite the court order directing the department to take remedial action has however proved to be a deterrent to any further action in this regard. Farmers in the tail ends comment that the inability of the court to enforce its orders had prevented them from renewing their attempts.

Clinging on to the private

Most of the conflicts and contestations cited above have arisen when attempts have been made to encroach into the public or common status of certain water sources. Conflicts also arise when the 'private status' of certain water sources is threatened. The conflict arises when right holders are unwilling to partake of their private rights to the resource concerned.

Stream water in the ponds

The contestations that emerged amongst the potential beneficiaries of the proposed Peringotukavu lift irrigation scheme indicates the prized nature of privately owned ponds. One group of farmers who protested were those who owned land in the ayacut area of the Karinkulam, which was one of the many ponds into which the pumped river water was to be stored. These farmers felt that by storing river water in their pond, they would have to share their water rights with farmers outside the pond ayacut, for the water stored in the pond was actually river water on which anybody could stake a claim. In other words, the water stored in the pond would become public property of all whose lands were located in the command area of the proposed Peringotukavu lift irrigation scheme. As Manikkan, a landowner in the command area of the Karinkulam remarked, if river water was to be stored in the pond, in no time one would find a number of pump sets all along the pond, and the pond would be dry in no time. Manikkan and other farmers observed that they were able to raise a crop of paddy many a time only because of the storage of water in the pond. Once this had to be shared with other farmers, they would lose the privileged status they currently enjoyed. Such schemes they felt would change the existing formulation of water rights from ponds, which was currently in their favour. These farmers therefore were of the opinion that even if the Peringotukavu lift irrigation scheme was to be sanctioned, they would not agree to the proposal of storing the river water in their ponds.

Conflicts within pond ayacuts

While vesting ownership rights over ponds with the hitherto cultivating tenants, there was little clarity on how rights to water were to be transferred in the event of the future sale of land. Similarly, the rights of pond owners, vis a vis the owners of the land in the pond ayacut were also not clearly stated. The absence of clear rules and procedures in this regard has resulted in a situation wherein landowners can choose not to give water rights to purchasers of land in the ayacut of a pond.

The legal basis of water rights lies in the title deed that pertains to each land holding. This mentions the rights of the landowner with respect to the extent of land and the rights to water. A title deed pertaining to a piece of land that comprised a two-fifth portion of the pond ayacut, would mention that the owner of that holding had a two-fifth right to the water in the pond as well. Prior to the introduction of the canals and the energized pumping devices, the title deed would also mention the kind of water rights associated with the land. In the case of single cropped land, it would be mentioned that water could be manually lifted out from the pond only during the first crop season. At the time of land reforms, some of the tenants purchased ownership rights over the pond as well, by paying an additional sum of money to the Land Tribunal. Since an additional sum of money had to be paid, it was mostly the large tenants who purchased such rights, which was then mentioned in the title deed. Hence, amongst the landowners in a pond ayacut, only some have ownership rights over the pond per se. This implies that they have ownership rights over the trees on pond bunds, over the fish in the pond, and even to the silt deposited in the pond bed. Currently when ponds are being deepened, the soil that is removed is often sold to outsiders, with the pond owners having a right to the sale amount. While all landowners in a pond ayacut are entitled to the water stored in the pond (proportionate to their share of land in the ayacut), pond owners assume that they have superior rights over non-pond owners.

Conflicts in this realm have reportedly increased with the increase in the number of sale and purchase deeds in the years following the reforms. In the event of sale, the seller is expected to transfer all land and water rights that pertain to the concerned holding to the purchaser.[3] However, when owners of entire pond ayacuts sell a portion of the land, they are often disinclined to give away their ownership rights to the pond. So the ponds often remain in the ownership of the old owners, even when parts of the ayacut have been sold off. Similarly, in certain cases, the seller is found to withhold even rights to the water stored in the pond to the purchaser. This is usually the case when the seller is selling only a part of his individual holding or a part of a larger family holding. In such cases, he is often disinclined to give water rights to the purchaser in order to prevent further fragmentation of the existing stock of water. This leads to a situation wherein a portion of the pond ayacut which till then enjoyed water rights is deprived of water from the pond. Buyers react in two ways. Some of them insist that water rights be given in writing to them. In such cases they are not cheated of their rights. In some other cases, farmers are not so vigilant, assuming that purchase of land will automatically confer upon them rights to water. This is particularly the case when the buyer is not from the immediate locality, and is therefore not aware of the intricacies of pond management and pond water rights. In such cases, the buyer is not aware of the possibility of being cheated. In some cases farmers do not insist on being granted water rights as they assume that the water through the canal system will be adequate. Many conflicts are resolved by referring to older title deeds that pertain to the same plot of land, referred to as prior deed, or parent document in legal parlance. In local parlance, they are referred to as 'adi adharam' (adharam meaning title deed), which mentions clearly the water rights of the particular piece of land. Hence even if the seller does not confer water rights, it can be restored by referring to the prior title deed which can be retrieved from the Sub-Registrar's Office.

Take the case of Swaminathan who purchased 3.2 ha of land in the Matacode area from a prominent land owning family in the area. Despite his land carrying water rights to the pond, he was not given rights to the water. As a result, when he attempted to take water, he was stalled by the other right holders. Rather than contest the issue he thought it better to invest in a tubewell. The reason he cited was that being new to the area, it was difficult for him to mobilize local opinion in his favour. The tubewell however did not yield adequate amounts of water. When the supply through the canal system was inadequate, he would suffer crop losses while others in the pond ayacut managed to hold on. After incurring repeated crop losses, he sold his land and house to another large farmer in the area and left Kollengode. Aziz shared a somewhat similar fate. He purchased land in the ayacut of the Odungatuchira pond, which has been surrendered by the former landlord as 'excess land' during the implementation of the land reforms. When a pond is surrendered to the government, then the right to use of water stored in the pond is not confined to the landowners in the ayacut, but is open to the wider pubic. This was not disclosed to him, and he assumed that he would be able to take water from the pond. During one cropping season Aziz faced a water crisis and pumped water from the pond. This led to a public protest by the washer folk in the area (see p. 248 for details of the conflict), and since then Aziz has not been able to take water from the pond. In this case, Aziz was not aware of the rights situation before purchasing the land.

In certain rare cases, even the parent document has been manipulated. Hamid's case deserves attention in this regard. Hamid invested all his savings in buying a 0.8 ha holding in the ayacut of the Velanganpadam kulam. He had purchased a part of the share of one of the three siblings who had inherited the Velanganpadam kulam ayacut from their father. He was given water rights in writing. During the first few years he took water from the pond, and even pumped water when needed. He was even given a share of the fish in the pond. However, he soon entered into a difference of opinion with the seller of the land who continued to own land in the same ayacut. The seller told Hamid that he was not entitled to the water in the pond, for as per the original title deed, his land was not entitled to water. On cross checking the document, Hamid found this to be true. This was a fall out of an earlier dispute between the former landowners, as a result of which one of them had been deprived of water rights for certain fields when the title deed was written. In reality, Hamid's plot of land was a part of the pond ayacut. This is borne by the fact that when an earlier tenant managed the entire pond ayacut in the pre land reform era, he used to take water to what are Hamid's fields today. The agricultural workers who had been working on his land for the past two generations also recall that water had been taken to this piece of land from the pond in the past. Hamid had been clearly cheated, for though he was given water rights in writing at the time of land purchase, his claims to water would not stand in court, as the original title deed did not confer water rights to his plot of land. The case was taken up in the local '*naatukootam*' (an informal gathering of local people). None of the other landowners in the ayacut supported Hamid. Hamid feels that since his plot of land was customarily entitled to water, his case may have legal standing, despite the existence of a title deed that does not support his claim. However, the cost of filing a case in court and fighting it out prevented him from taking legal recourse. One of the landowners in the pond ayacut summarized the situation as follows:

From the angle of justice Hamid is entitled to water, but in the eyes of the law, he is not as the title deed stands against him.

As a result, when all the landowners in the pond ayacut raise a crop of paddy, Hamid loses his, as he has to wait for the canal water to reach him on time.

One of the ongoing conflicts during the period of fieldwork arose as the owner of the pond ayacut sold the land in the pond ayacut and the pond to two separate individuals. In 1989, the owner of the Tekke Eri planned to sell the paddy land in the ayacut of the Tekke Eri to Kumaran and the pond per se covering an area of 0.4 hectare and the adjoining parambu land covering 2.4 hectares to another person by the name of Rajan. In order to safeguard his rights to water in the pond, Kumaran offered to purchase the pond as well, for a total cost of Rs 9,500 but the owner of the pond had already committed to sell the pond to Rajan. Kumaran however ensured that the sale deed explicitly mentioned his entitlement to the water stored in the pond. The sale deed also mentioned that in recognition of his water rights in the Tekke Eri, under no circumstances, could the pond be filled up or destroyed. A couple of years later, Rajan, the new owner of the pond, agreed to sell the pond and the adjoining parambu land to the Kerala State Electricity Board (KSEB) who wished to set up a sub-station at the site. The KSEB planned to fill up the Tekke eri. Kumaran came to know of this move and requested the KSEB not to go ahead with the pond filling and the construction. The KSEB responded by saying that they were acting as per government orders and that if Kumaran wished to stall the filling of the pond, he should file a case in the court. Kumaran filed a case in the court, against the District Collector, the Chief Engineer of the Electricity Department and the Tashildaar. The KSEB stopped all further construction work till the case was resolved. The court decreed that the pond should not be filled or destroyed, and that it should be left intact. Kumaran in this case had to spend Rs 30,000 towards court fees, lawyer fees and other incidental charges. Pond filling of this nature not only harms the interests of the farmers whose lands are located immediately below (like Kumaran's) but also the interests of farmers whose lands are located further downstream, which benefit from the seepage of water stored in the pond above.

The emerging conflicts over ponds indicate the complicated legal context within which pond rights are located. They illustrate how landowners manipulate sale deeds by false statements regarding water rights. A detailed study of some of the conflicts, along with a study of the title deeds and other documents would yield more insightful details. This has not been possible within the time frame of this study. These conflicts indicate that the urge to privatize is prevalent at all levels. While storing public canal water in private ponds represents one level of privatisation, not sharing the pond or rights to the water in the pond with those who own land in its ayacut represents a more acute form of privatization. These conflicts indicate how important the private status of the ponds is to the farmers who benefit from the existing formulation of rights. It is no surprise therefore that farmers are unwilling to surrender their ponds to the panchayat, even if it implies that their silted up ponds would be desilted at the panchayat's expense. These conflicts also point to the importance of vesting the ownership rights over ponds with a common body, and giving landowners only usufruct rights to water.

Private Versus Common Good

The situation is best described as muddled. While each water source is classified under a particular property regime, the mixing of waters, if such a term exists, confuses this classification. In some cases this mixing is not engineered, as in the case of water distributed through the canals seeping into the streams and rivers. In some cases it is, as in the case of canal water being diverted into ponds, or water from streams lifted into ponds, leading to contestations. In such a maze, those farmers with access to multiple sources of irrigation are comfortably placed. They are able to use an interlinked system of irrigation, and compensate the shortages in one with the surplus in the other. This has implications on both equity and sustainability. While privately controlled water sources lead to unequal access, it also allows private owners to extract water without restraint, thereby leading to an unsustainable pattern of water consumption. As Bhatia puts it in the case of groundwater, growing inequity in the use of the resource is both a critical consequence and a major cause of over exploitation (Bhatia 1992).

Then there is the issue of technology. The availability of energized pumping devices strengthens private control over water. This not only pertains to groundwater, but also to surface water. Farmers with the ability to invest in costly pumping facilities are able to appropriate larger amounts of water from a common pool such as a stream and store it in their private ponds. On the whole, tubewell owners, riparian land owners and single pond owners are at a distinct advantage for they are able to take advantage of access to privately owned water sources and the availability of energized water lifting devices. Needless to say, the absence of any kind of restriction on the total amount of water extracted comes to their aid. Once again, along with equity, sustainability is also at stake.

Ensuring greater equity and sustainability in the use of water requires a reconsideration of existing property classifications over water. Simultaneously, the role of modern energised lifting in promoting the unrestrained extraction of water also needs to be subject to greater scrutiny. Given the predominantly private mode of appropriation, the main problem as Benda-Beckmann puts it, lies in a combination of an unrestrained individual right of appropriation with individual ownership rights over the means of appropriation (Benda-Beckmann 1992), in this case energized water lifting technology.

Reconsidering existing property classifications

During the past two decades, it is being increasingly recognized that an unregulated private regime encourages indiscriminate consumptive utilization of water and leads to rapid depletion (Torori et al. 1996). Private property rules have also been critiqued for treating natural resources as temporally and spatially bounded commodities (Ojwang and Juma 1996). Excessive exploitation of water in particular has led to an emphasis on alternative forms of property arrangements over water. While most of the alternatives suggested pertain to groundwater use, which has been firmly rooted in the private domain, surface water sources are also being subject to such reconsiderations.

In the case of groundwater, the present property rights structure is one that considers the individual landowner's right to appropriation as sacrosanct. It has therefore been critiqued for providing incentives for over exploitation of the resource (Bhatia 1992). Bhatia suggests that some form of collective rights over groundwater should be defined as an alternative to the prevailing private mode, but admits that there is little clarity on how an alternative property regime should be defined. Meinzen-Dick also argues in favour of community or jointly owned tubewells in order to ensure greater equity (Meinzen-Dick 2000). Bhatia argues that the definition of collective rights to groundwater should have legislative backing. Legislative changes in this direction she argues would bring changes in social perceptions about what constitutes 'legitimate' use of the resource. Bringing about a change in social perception has been emphasized by others as well. Burke et al argue for instance that trying to exercise a measure of equity and control over the abstraction and protection of groundwater is thwarted by the perception and treatment of groundwater as a privately owned resource (Burke, Sauveplane and Moench 1999: 311).

In the case of surface water too, there have been suggestions to reduce the exclusive use rights granted to riparian land owners. The law of riparianism has been critiqued for confining the right to the use of running water to those who own the land abutting a stream or river (Singh 1992, Orindi and Huggins 2005). This has been illustrated in the study area as well, wherein farmers who do not own riparian land do not have the right to dig channels through the riparian land owned by other farmers in order to irrigate their land. Instead, they have to irrigate the formers land before they can irrigate their own. In this context, Orindi and Huggins argue that public access points or rights should be provided to those who do not own riparian land, in order to prevent such exclusion from being passed over to future generations (Orindi and Huggins 2005).

Similar suggestions have emerged from farmers in the study area as well. Farmers without access to ponds in the study area have argued that all ponds should be declared as government property and that all farmers should have a right to the water in ponds. Interestingly Bhatia reports of small and marginal farmers in Gujarat coming up with a similar suggestion as far as ownership of tubewells is concerned (Bhatia 1992).[4] In the case of ponds I would take the argument further that it is not only irrigation that merits consideration as far as the water stored in the pond is concerned. The livelihoods of the lotus tuber pickers, the washer folk and the fisher folk should also merit consideration while formulating rights to the use of water in the pond.

There is therefore an emerging consensus that exclusive ownership and use rights are not in the interests of equity. Community or collective ownership however carries its own problems. Power differences within communities are known to topple down any objective of equity and transparency in decision-making (Meinzen-Dick and Zwarteveen 2001) as well as in resource sharing. At the same time, the relevance of collective systems of ownership and use cannot be ruled out. Upadhyay while emphasising the relevance of group rights in water management in order to resolve existing and potential conflicts surrounding the access and control over water resources, underlines the need of a better appreciation of internal water rights within the group, arguing that it is important to lay down the right of the group members vis a vis one another (Upadhyay 2002).

The relevance of a 'water ceiling'[5]

Reduction in present demand levels is considered an important strategy in abating the water crisis, even though there would still be problems of existing levels of resource conflicts and environmental degradation (Rogers and Hall 2003). Existing legal traditions have however proved to be ineffective in regulating the use of modern technologies, particularly since the advent of modern drilling techniques and powered pumps (Vincent 1990:12, Hodgson 2004). So along with a reconsideration of the existing legal categories in such a way that individual rights of appropriation are curtailed, we also need a regulation of the use of modern technologies.

While the imposition of land ceilings has been an accepted method to prevent concentration of landholdings, a similar concept in water use could help to restrict uncontrolled exploitation of water. Orindi and Huggins argue that fixing an upper limit on individual extractions are an important means of regulating the withdrawal of water given the present day proliferation of individually operated pumps (Orindi and Huggins 2005). Van Koppen et al. suggest that a water tax be imposed on large scale water users, which would not only help in financing water management services and generate net income, but would also provide the incentive to use water prudently (Van Koppen et al. 2004). Water charges/fee and water permits are some of the other suggested measures. The above-mentioned measures form the crux of water reform initiatives undertaken in many parts of the world (Van Koppen et al. 2004, Sokile and Van Koppen 2003).

In the context of groundwater, restricted power supply and increases in tariff rates are some of the suggested measures to control water consumption (Saleth 2005). Bhatia suggests the banning of institutional credit for the purchase of water extraction mechanism from all water sources in areas of severe groundwater scarcity, expect in the case of small and marginal farmers; introducing a system of electricity pricing wherein the consumer pays for each unit consumed in addition to the flat rate paid annually for each electric motor. She also recommends a system whereby higher electricity charges are levied in areas where it is important to discourage the consumption of water, and also that the flat rate charge be fixed differently for different users, by linking the horsepower to the landholding. All these suggested regulations are applicable to surface water extraction as well. Such regulations could help to bring some control over the unrestrained exploitation of water from streams, ponds, and wells in the study area.

All of the above measures assume some form of external enforcement and monitoring. In addition, regulations per se have little value. In the case of groundwater consumption for instance, while restricted power supply is considered to facilitate some sort of regulation, its effectiveness has been undermined by the ability of the rich farmers to invest in large pumps and multiple wells, as well as by availability of diesel pump sets (Saleth 2005). Voluntary compliance with such regulations therefore is critical in ensuring their effectiveness. There is evidence of a number of organisations that impose constraints and monitoring upon themselves, that have shown themselves to be effective in managing water resources (Sick 2007). Formulation and enforcement of such measures at the community level has also been advocated for providing opportunities to instill a sense of responsibility for the conservation of community resources (Bhatia 1992).

This is an issue that has been consistently sidelined. While the Indian Water Policy recognised the need to limit individual and collective water withdrawals, it has failed to identify the institutional mechanisms required for defining and enforcing such physical limits (Saleth 2005). In the study area, the institutional vacuum that exists at the level of individual water use is a significant one. Despite the existence of the karshaka (CADA) samitis, the beneficiary groups constituted under the Jalanidhi programme,[6] watershed user groups, as well as bodies such as the Project Advisory committees, there is no effective mechanism in place to regulate the current pattern of water consumption. There are no bodies to facilitate joint decision-making regarding water distribution and water use. Equally significant is the notable absence of attempts at collective action to redress the inequities discussed above. Even after the implementation of decentralisation reforms, there is no organisation at the local level that addresses the sustainability and equity dimensions of water resources management in the study area.

Conclusion

This chapter has illustrated how the existing, often shifting property classifications over different irrigation sources create inequity in access to water. In particular, it has discussed how the classification of ponds and wells as private property enables private (mostly individual) owners to carve out private enclosures of water. This is well illustrated in the interlinked system of ponds, tubewells, canals and streams that some of the resource rich farmers have set up. The existing property rights system, wherein water rights are defined by land rights, legitimises such enclosures.

The chapter also discusses how property categorisations, by being disregardful of the hydrological interconnectedness between different sources of water, promote both unsustainable and inequitable use of water. This is accentuated by the property-technology interface, wherein the use of energised lifting devices facilitates unrestrained use of water, particularly from privately owned water sources. Apart from the negative implications on long-term sustainability, it also widens the gap between the resource rich and the resource poor farmers. This is manifest in the higher incidence of crop failures and poor yields amongst the small and marginal farmers, as compared to the large farmers in the area.

Conflicts over the property status of water discussed in the chapter illustrate the tendency to privatise common flows, whether through the canals or the stream or in ponds. Conflicts within pond ayacuts illustrate how the tendency to privatise is prevalent even at the most micro level. It is also important to note that in the absence of any corrective action, such attempts at privatisation are unquestioned. Even in the panchayat (public) ponds, public protests were required in order to restrain the operation of private interests. This shows the level of regulation and vigilance required to ensure that certain water sources are maintained as public or common.

Notes

1. In May 2002 I met a group of tuber pickers at work at the Choorikad kulam. About thirty of them had left Pallasana in the morning, in teams of five, and each went in a different direction.
2. The procedure is as follows: the farmer who wises to install an electric pump set to lift water from a private water source such as a pond or well, renders an application to the office of the Agriculture Department, i.e. the Krishi Bhavan. The Agricultural Officer either visits the site or deputes another official to do so. The official concerned is expected to make a recommendation for the installation of the pump set after inspecting the water source concerned and on assessing the extent of land to be irrigated. The recommendation that

the farmer concerned be allowed to use a pump set of a particular horse power is made after assessing the water requirements of the land to be irrigated. This recommendation is then forwarded to the Electricity Department, which then gives the sanction for the purchase of the pump set. Farmers report that in getting the sanction for the pump set they have had to bribe officials in both these departments. When farmers have not bribed the officers in some cases the sanction for the pump set was delayed without any reason.

3 Conflicts over transfer or non transfer of water rights along with the land in the event of sale of land have been reported from Nepal. See Sodemba and Pradhan 2000 in Benda-Beckmann and Benda-Beckmann 2006).

4 Bhatia observes that small and marginal farmers who bore the brunt of over exploitation of groundwater by the resource rich farmers, suggested that the government install bore wells for public use and prohibit the functioning of private tubewells (Bhatia 1992).

5 Van Koppen suggests that the imposition of water ceilings can curtail the water rights of the resource rich sections of society (Van Koppen 1999:13).

6 A government programme for rural water supply and sanitation, implemented throughout the state with aid from the World Bank, since 2001.

7

An Outline for a Water Reform

Through an enquiry into the phenomenon of agricultural water scarcity in Palakkad, this book has shown how institutional, technological and ecological (natural) factors play together to shape the management and distribution of water resources. The focus has been on understanding the implication of this interplay of institutional, technological and ecological dimensions on ecological sustainability and social equity, and thereby on the current manifestation of water scarcity. Before going into the main conclusions, I briefly present the three main fields of enquiry adopted by the study:

1. To review developmental policies, particularly irrigation and agricultural policies, and the technologies that have given shape to the present style of water resources management. This includes an analysis of supply-oriented water development approaches, focused on a single mode of irrigation that has led to the neglect of irrigation and agricultural practices that were shaped by the agro climatic and topographic features of each region. Also analysed is the hydrological impact of the promotion of water intensive cropping patterns, viz. paddy. This has been dealt with in Chapters 2 and 3.
2. To understand how the prevailing mode of resource appropriation and the increasing privatisation of rights over resources and technologies affect the distribution of existing supplies of water. More specifically, to understand the sustainability and equity implications of increasing privatization of a common-pool resource. This has been dealt with in Chapters 4 to 6. While Chapter 4 discussed the strengthening of a private property regime over land and water in the post land reform era, Chapters

5 and 6 focused specifically on the extending private domain at the cost of a shrinking public/common pool of water.

3. To present the growing ecological fragility of the local water resource base and the increasingly inequitable access to it, and to thereby illustrate the inadequacy of supply-oriented technological measures in abating water scarcity. To understand the role public agencies and local institutions have played and can play in shaping a pattern of water resources management that is cognizant of the dimensions of both sustainability and equity. This has been discussed in Chapter 3.

Detailed interviews with farmers on changing land and water management practices, observation of water management and distribution practices in the canal and the pond systems, of emerging conflicts over rights to water, along with a detailed perusal of secondary literature provided the material for the study. The running argument in the book has been that the technological and institutional dimensions of the prevailing mode of water resources management have not been adequately cognizant of the ecological properties of water resources and therefore to the requirements of sustainable management. Equity in distribution of water has also not been prioritised. I have therefore argued that the water crisis of the study area can be addressed only if technological and institutional designs cater to the needs of sustainability and equity. The following sections summarise the main findings that have emerged from the three broad realms of enquiry outlined above.

Reviewing Irrigation Policies: The Sustainability Dimension

The irrigation policy of the state has been based on the assumption that the reservoir based canal irrigation mode was suited to all parts of the state, irrespective of regional variations in topography, soil characteristics, rainfall and cropping patterns. The focus on a single mode of irrigation was accompanied by the promotion of a single crop, i.e. paddy in all the irrigated commands of the state. However, as Chapter 2 has illustrated, most of the canal irrigation projects have not been functioning to their intended capacity, and the area irrigated by government canals has been on the decline in the state. Area under paddy cultivation has also been declining in the state. The situation is particularly worrisome in Palakkad as it has registered a decline on both these fronts, despite having received the maximum amount of irrigation investment.

It is therefore timely to re consider the irrigation and agricultural policies that we have followed so far. The poor functioning of the canal systems and the siltation of most of the large reservoirs in the state had alerted attention to the possibility of alternative forms of irrigation since the 1980s. Critiques of this irrigation policy have pointed to the relevance of smaller irrigation structures such as catchment based minor storages that take into account local agro climatic and geological features (Santhakumar and Rajagopal 1993, Kannan and Pushpangadhan 1988, Narayana and Nair 1983). They have also argued that local specific factors such as crop-mix, soil characteristics, crop specific water demands, topography, soil erosion and siltation problems should be considered before assessing the total water requirement of a region and the suitability of a particular mode of irrigation. In particular, the paddy dominated cropping pattern has been critiqued for necessitating the construction of large reservoirs. The need for a greater appreciation of locally generated water supplies has been emphasized, particularly keeping in the mind the run off that is generated on undulating terrain that characterizes most parts of the state.

Such critiques however have made little impact on the overall orientation towards irrigation and agriculture in the state. The changing irrigation strategy of the state reflected in a greater emphasis on minor irrigation (both surface and groundwater) since the late 1980s needs to be viewed with a pinch of salt. For one, the declining emphasis on major irrigation does not indicate increasing awareness of the adverse ecological implications of this approach. It has more to do with the financial implications of major irrigation development, and the exhaustion of topographic possibilities for the construction of large reservoirs (GOK 2003). While minor irrigation technologies are definitely smaller in scale, they do not take into consideration the above-mentioned factors such as crop mix, soil characteristics or topographical variations while supplying water. The ecological implications of such irrigation measures have also not been addressed. The cumulative impact of lift irrigation on the total available water supply in each stream remains unaccounted. Reports of falling groundwater levels from many parts of the state, including the Chittur Block panchayat of Palakkad indicate that groundwater exploitation is not a sustainable solution either. In addition, the emphasis on minor irrigation has not been the result of a reconsideration of currently water intensive cropping patterns in irrigation commands. To the contrary, minor irrigation has been advocated as it suits the water requirements of perennial crops such as coconut or plantain, which are not water conserving by any measure (GOK 2002a).

Despite the problem of water scarcity in various parts of the state, there is yet to emerge an irrigation policy that addresses the long-term sustainability of land and water resources. Irrigation is still seen to be far removed from concerns of water conservation. Currently, it is watershed management programmes that are considered to work towards the goal of water conservation. These programmes however can achieve little if they are not integrated with the mainstay of irrigation and agricultural programmes being undertaken across the state. Undertaking measures for in situ moisture conservation for instance will have little impact if no restrictions are laid on the extraction of water for irrigation by downstream farmers. It is therefore imperative that irrigation planning should be centred on the goal of ensuring the sustainability of water resources. In the context of canal irrigation reform for instance Mollinga has argued that in order to ensure ecological sustainability, irrigation should be viewed as a part of a larger hydrological system (Mollinga 2000). While sustainability concerns are being recognised, a conceptual framework that could provide a clear operational alternative to the present one is yet to emerge (Bandyopadhyay 2006).

Reviewing the paddy focus

An important component of an irrigation policy that promotes sustainability is the adoption of water conserving cropping patterns.

As discussed in Chapter 2, the state's promotion of large scale irrigation and the cultivation of paddy in all irrigated commands were in response to the need of achieving self-sufficiency in food supplies. In order to increase the area and production of paddy in the state, double and even triple cropping of paddy was advocated in all the irrigation projects of the state. In the study area, this resulted in the monoculture of paddy and the abandoning of cropping patterns, which were suited to the agro climatic conditions of the area. The cultivation of dry crops on single cropped paddy land was given up, so also was the traditional topographic classification of agricultural land into kalayi and potta.

The growing unreliability in the supply of water from the Chulliar reservoir is however making double cropping of paddy a difficult proposition. This is particularly the case with areas located in the tail ends of the main and distributary canals. It has therefore illustrated the lack of wisdom in altering cropping patterns merely on the assurance of externally supplied water. Despite problems with water availability, cultivation of less water intensive crops on single cropped land is not being experimented with. While individual farmers may experiment with a crop of black gram or groundnut, it is not the norm, as it was in the past. To the contrary, most farmers consider it shameful to experiment with dry crops on paddy land.

While dry cropping is not practiced on single cropped paddy lands, low lying double cropped paddy lands are being cultivated with non-paddy, water intensive, perennial crops such as plantains and coconut. They are also being converted into building sites. This is reflected in the state and district level data on the declining share of paddy in the crop-wise gross area irrigated, and the increasing share of crops such as coconut and banana. Hence, the decline in paddy area in the state and in the district of Palakkad cannot be treated as an indication of reduced water consumption. To the contrary, the spread of cultivation of perennial crops could as well indicate an increase in the water withdrawn from water sources such as rivers, streams, ponds and wells. Once again, the increasing share of pumping from wells, tubewells and streams, vis a vis the decreasing share of government canals in the net area irrigated in the district and the state points to such a trend. This can have serious implications, especially during the summer months, when the flow in streams and ponds is at their lowest. The situation is even more worrisome, as there exists no quantitative estimate of the total amount of water abstracted, and whether this is within sustainable limits.

Missing in the debate over irrigation and paddy cultivation in the state is the relevance of water efficient cropping patterns, from the point of view of water conservation as well as food self-sufficiency. The conversion of paddy land for the cultivation of non-paddy cash crops undermines the objective of securing self-sufficiency in food. In the case of Palakkad district as well as the study area, little attention is being paid to the cultivation of water efficient dry cereals. This is despite the fact that the district has a long history of dry crop cultivation. A region wise appraisal of total water availability and the suitability of the existing cropping patterns needs to be taken up in order to ensure sustainable levels of water consumption (Santhakumar and Nair 1999). This is particularly important given the increasing desiccation of the Bharathapuzha river basin in which the study area is located.

The poor state of the art of canal irrigation

The poor state of art of the canal irrigation infrastructure and the distribution systems is yet another reason why irrigation policies merit reconsideration. A thorough review of the functioning of all the irrigation projects in the Bharathapuzha river basin in Palakkad district would throw open important issues for debate regarding the future of canal irrigation in the state. This is particularly so as the district of Palakkad has been the forerunner of canal irrigation in the state. The construction of large reservoirs and the associated distribution networks have resulted in financial investment of a very high order. The construction of almost all canal irrigation projects in the state have entailed significant cost escalations, with Kerala recording the highest cost escalation amongst the Indian states (Vishwanathan 2002). These financial implications assume added seriousness given the fact that most of the irrigation projects in the state irrigate an area far less than was originally intended. The Gayatri

Project for instance was irrigating only 64% of the area originally intended as of 1995–96 (ibid.).

In addition, the distribution system is in disarray. Chapter 5 has discussed the distribution system in the Gayatri project. The missing shutters, locks and keys along with the enlarged outlets make a mockery of the intended system of water distribution. The infrastructure in disrepair is not only an indication of departmental inertia in correcting the situation, but is also an indication of power plays along the canals, both above and below the outlet. This is manifest in shutters being removed at will, and locks and keys in the custody of farmers and not maestries. The disinclination of the irrigation department to improve the distribution system is also reflected in the fact that canals do not get lined and field channels do not get extended despite repeated requests. This is despite the fact that the farmers of the second reach had filed a case against the department in the court, holding them responsible for non-availability of water to the second reach.

Present day canal irrigation reform, commonly referred to as Participatory Irrigation Management (PIM) is focused on the constitution of water user associations who are to take over management tasks from the irrigation department, particularly with regard to water distribution. The constitution of such user associations has commenced in the adjoining Malampuzha Irrigation Project. Once again, such a view of reform ignores the more deep-seated problems discussed above. Mere constitution of water user associations without addressing the skewed power relations amongst farmers in each association or the corruption that prevails over the functioning of the irrigation department will achieve little. While lining the canals and field channels and restoring outlets to their original dimensions is important, such measures can lead to an equitable distribution of water only if the above mentioned factors are addressed.

There seems to be little debate within the irrigation department on how to reform the situation. During the past two decades, implementation of the Palakapandi diversion scheme (see Chapter 3) has been portrayed as the sole solution to the water crisis that the region faces, thereby obscuring the severe anomalies in the existing distribution system. The important question as to how the additional water supplied through the Palakapandi project is going to be distributed is ignored.

Local level water resources management planning

The ecological ill effects of the supply oriented approach to irrigation and the dwindling supplies of water have led to a growing emphasis on local level water harvesting and conservation measures. Caution has been expressed against being hypnotized by visions of long distance water transfers (Iyer 2001) and an excessive reliance on external supply of water which results in further neglect of local level ameliorative measures (Reddy 1999, Mehta 2004, Bharwada and Mahajan 2002). The study has also illustrated how the excessive reliance on external water supply has reduced the local irrigation sources (the ponds and the streams) into appendages of the canal system. In addition, the increased access to surface and groundwater through energised lifting devices has created a false image of local water availability that takes little consideration of physical limits to extraction. As Burke points out, engineering efficiency has sheltered the user from hydrological reality and the inherent limits of land and soil-water systems (Burke 2000).

If we are to prioritise on local water supplies then regional water resources management plans that focus on the regeneration of local water resources need to emerge. Devising locally appropriate water resources management plans has received some amount of emphasis in the state since the implementation of democratic decentralisation. This book has however illustrated that despite vesting the panchayats with control over water resources within their jurisdiction, there was no significant change in the overall approach towards water resources, which continued to be oriented towards enhancing the extraction of water, and not towards ensuring sustainability of the water sources (see Chapter 3). Neither has it contributed towards developing a regional understanding of water resources management and the problems therein.

It is significant therefore that the present Eleventh Five Year Plan proposes that all development activities undertaken by a panchayat, particularly in the realm of agriculture and irrigation, fall in line with a larger watershed plan. These watershed plans which are to be prepared by the concerned panchayats are to be part of larger river basin plans as well. In addition, the panchayats have also been entrusted with the implementation of the National Rural Employment Guarantee Act (NREGA) since 2006 (http://nrega.nic.in). This act aims at the generation of rural labour opportunities through the regeneration of livelihood support systems, particularly land and water resources. Afforestation and protection of local water resources figure prominently in the list of activities that can be undertaken by the panchayats in the implementation of the NREGA.

Effective implementation of the Eleventh Plan guidelines as well as the NREGA pre requisites a comprehensive regional understanding of issues related to water and natural resources management, reasons for degradation and long term ameliorative measures. It also calls for an appreciation of the relevance of watershed based restorative processes that ensure the sustainability of land and water resources. This calls for a fundamental change in approach towards the management of water resources, requiring considerable capacity building at the panchayat level. While the plan guidelines give in detail the institutional blue print necessary for such a transformation (such as constitution of special 'watershed sabhas' that discuss and deliberate upon watershed based land and water management), it does not mention how such a transformation in approach is to come about. This is particularly important as the experience with decentralisation during the past decade indicates that there has been no significant change in the underlying approach towards water resources management. Even in the realm of water conservation, the emphasis continues to be on structural measures such as check dams. Afforestation and vegetative measures that enhance soil water recharge have been sidelined.

Any possible integration of irrigation and agricultural activities with that of sustainable land and water management would imply that water extraction for irrigation is subject to considerations that ensure the long term sustainability of the water source in question, be it a pond or a stream or a well. These plans should also deliberate upon the kind of cropping patterns that are most suited to the local agro climatic features. As far as the study area is concerned, the ponds that tap into locally generated run off, and the earlier classification of agricultural land into kalayi and potta begins to assume relevance.

While the present plan guidelines mark a significant shift in mainstream thinking on land and water resources management, it is important to take note of the fact that such an integrated vision is being attempted only in the realm of panchayat level planning. It is only 30% of the total fund allocation of the state government that has been devolved to the panchayats. Hence, majority of state spending continues to be vested with bureaucratically organized government departments, who continue with the prevailing supply-side approach. It remains to be seen whether state level irrigation and water resources planning takes into consideration the watershed plans prepared for each catchment by the panchayat concerned. At present, while decentralisation takes place at one end, at the other end, the irrigation bureaucracy continues to draw plans for inter basin transfers of water and the construction of reservoirs wherever possible, despite the fact that they do not adhere to any 'catchment plan' or catchment logic. This has been illustrated in the study area where the scarcity problem is sought to be resolved through the implementation of inter basin transfers of water, and not through any local level ameliorative measures.

Water and Property

Chapters 4–6 of this book have dealt with the relationship between property rights regimes over land and water, and the management and distribution of these resources. The book has illustrated how the existing property rights regime, by facilitating unrestrained access to certain water sources on one hand, and skewed access on the other hand, results in an unsustainable and inequitable mode of water resources management. The main issues that emerge from this analysis are

a) Unrestrained and skewed access that a private property regime facilitates.
b) The framing of property rights without any reference to environmental sustainability, leading to unsustainable use patterns.
c) A reconsideration of existing property classifications and the desirability of collective ownership systems.

Ecology in property rights

This book has illustrated how the existing formulation of property rights, by being rooted in the superiority of human entitlements has paid little attention to the need of long-term protection of the resource concerned (Klug 2002, Freyfogle 1996). In the case of water, since most of the water sources are either under state control or private control, the focus has been on defining the rights of the state or the individual right holder, and not on the resource per se. Freyfogle has explained this dimension in the case of land rights, pointing out that rights to a parcel of land are not framed with respect to the specific features of that parcel of land, i.e., with respect to the soils, terrain and vegetation (Freyfogle 1996). As a result, the land use options pursued by the landowner need not take into consideration the natural features of the land, and can thereby result in ecologically unsustainable land use patterns, as has been illustrated in the case of conversion of paddy lands in the state and in the study area. Disregard for the ecological features of the land was also demonstrated during the process of reorganization of rights to land as a part of the land reform process initiated in the state during the 1970s. The interconnections between different land use categories (i.e., parambu land, paddy land and forest land) were ignored in the process of physical fragmentation of the land.

Existing property categorisations have also been critiqued for their inadequate understanding of the ecological properties of the resources concerned (Klug 2002). This is particularly so in the case of an extensive and fluid resource like water which follows no clearly defined boundaries (Sick 2007, Hodgson 2004). The need for a property rights regime that recognizes the physical features of the resource has been emphasized in the case of groundwater (Burke et al. 1999, Bhatia 1992). The exclusively private property mode of ownership that prevails over groundwater has been viewed as unsuitable to the management of a resource that does not follow land ownership boundaries (Meinzen-Dick 2000; Burke 1999). This is true of all other water sources as well. While water stored in a pond may seem to be a confined resource and hence can be more easily classified as a privately owned resource, the water stored in the pond does interact with the groundwater table. Hence excessive extraction of water from a pond can lower water levels in adjoining wells and vice versa. Similarly, water abstraction from streams can threaten water levels in wells that draw upon the same aquifer and vice versa (Glennon 2002, Schlager 2006). The formulation and operationalisation of water rights needs to be based on an understanding of the unique ecological properties of the resource. This has led to the argument that being hydrologically interconnected, surface and groundwater systems need to be governed by a single property rights regime (Hodgson 2004).

Recognition of the sustainability implications of current resource use patterns manifest in polluted water and soils, depleting aquifers and drying streams has led many countries to formulate various kinds of regulations over resource use. These have taken the form of land use restraints and zoning procedures in the case of land, and the allocation of water permits and pollution regulations in the case of water. They attempt to curb the access to resources granted under the prevailing rights regimes. It is, however, recognized that while these regulations attempt to bring about some restraint over the use of resources, they do not alter the basic shape or character of the concept of ownership (Freyfogle 1996, Hodgson 2004). Property law continues to be focused around the superiority of the rights of the individual land owner, and violation of these regulations does not result in a confiscation of rights. In addition, such regulations can easily be overcome through back-door measures (particularly in the Indian context), and hence they have not reduced the absolute nature of rights that owners of land and water enjoy. Hence if sustainability in the use of resources is to be ensured, regulations alone will not redress the situation.

A durable solution to the problem of over-exploitation and inequity in the use of water resources will then have to concern itself with the current conceptualisation of property rights (Bhatia 1992). The nature of the resource in question should not be external to the formulation of rights, but internal to it. Rights to resources therefore need to be framed in ecological terms. In the case of water this implies that rights to water are framed in such a way that people begin to use water in ways that not only promote human economy but also the integrity of the hydrological cycle (Freyfolge 1996).

Enclosed waters

An outcome of the neglect of the ecological properties of water resources in the formulation of property rights is the existing categorization of water resources into public and private water bodies. While such a categorisation has been critiqued as 'hydrological nonsense' (Hodgson 2004), it continues to determine access to water in many parts of the world, resulting in skewed access to the resource. More important, it creates private enclosures of water. The interlinked system of privately owned ponds, wells, shallow pits, and the operation of private pump sets on streams and rivers is an illustration of the same.

In addition, rights to water in the study area are tied to land rights, and hence aggravate the inequities precipitated by unequal land holdings. In the study area, this has largely been a fall out of the implementation of land reforms. The land reform exercise has been critiqued for approaching the issue of inequality through land ownership alone, ignoring the important issue of access to water. Recent works have pointed to the importance of assessing the existing inequities in access to resources other than land, such as water, in order to address issues of rural poverty and livelihoods, particularly of the small and marginal farmers (Tiwary 2006).

The tying up of water rights to land rights and the resultant creation of private reserves of water has led to a situation wherein the majority of large landowners have access to multiple water sources. From the sample of farmers selected from three broad landholding categories, apart from access to canal water, 58% of large farmers have access to more than one privately owned source of irrigation. Only 8% of the medium framers and none of the small farmers enjoy such access. Being privately owned, these landowners enjoy unrestrained access to water in these sources. It is pertinent to note that 67% of large farmers have access to individually or family owned ponds, which are not shared with a wider group of farmers. Farmers with access to these privately owned ponds have the right to suck them dry, irrespective of its implication on the water availability in near by water sources. At the other end of the spectrum, 67% of small farmers do not have access to any water source other than the canal system, through which supply is highly unreliable.

The private property status ascribed to ponds plays a pivotal role in aggravating inequity. It enables the privatisation of non-private waters that flow through canals and streams. By storing this water in private ponds, farmers are able to create private reserves of water that is theoretically to be distributed in equal amounts to all farmers entitled to the same. It has also enabled some of the large farmers to set up an interlinked system of irrigation, wherein water diverted from canals and water pumped from tubewells and streams is stored in privately owned ponds, and then taken for irrigation. These farmers are therefore buffered against the unreliable supply of water through the canal system as well as from the deviations in the normal rainfall pattern. There have even been instances where the public water stored in private ponds has been later sold to those who needed it.

In addition to the practice of storing public flows in private ponds, the differential abilities of large and small farmers to invest in energized water lifting results in differentiated access to water sources that are considered to be public. Since the extraction of water through the use of these devices is not subject to any restriction, it further strengthens the private control of the wealthy farmers over water. While small farmers like Velan have to take the last reserves of kerosene that would otherwise be used as cooking fuel for pumping water, electric pumps owned by larger farmers like Ponnunni or Bhaskaran work continuously. Hence access to water in different property categories is further differentiated by the material and social conditions of access (Crow and Sultana 2002).

In a nutshell, a combination of individual rights to appropriation of water and individual ownership rights over the means of appropriation results in unrestrained access and use of what is actually a common pool resource (Benda-Beckmann 1992). Such a situation has been aptly described as the 'tragedy of individual ownership' (ibid.). The implications of such a trend can get severe when the existing supply of water is decreasing. The growing tendency to enclose as much water as possible within private land holdings, i.e. to create private water enclosures, is at the cost of common or public water rights. This has been illustrated even within pond ayacuts, wherein collective water rights are threatened by legal manipulations (Hamid's case in Chapter 6). Restricting private encroachments into commonly or collectively owned water resources, such as the streams, canals and ponds is therefore critical to ensuring a fair and just distribution of existing water supplies.

The possibilities of collective rights

In the present situation, if greater equity in access to water is to be ensured, the prevailing private property status attached to water sources such as ponds and wells will have to be reconsidered. The highly contested nature of private rights to water and its consequences on equitable distribution has led to the argument that collective rights to water need to be constituted (Benda-Beckmann and Benda-Beckmann 2003). In the case of groundwater it has been argued that a formulation of collective rights is essential to ensure some measure of equity (Meinzen-Dick 2000, Bhatia 1992). The same has been argued in the case of riparian rights as well (Orindi and Huggins 2005). Such arguments are bound to gain in strength with the increasing shortages in water supplies.

Collective rights have been advocated for common pool resources from the sustainability angle as well. The basic feature of such resources is that they cannot be temporally and spatially bound (Sick 2007). McKean notes that granting individual private rights to such natural resource systems leads to chopping up of these systems into environmentally inappropriate bits and pieces and awarding rights in bits to individuals (McKean 2000). This is true of the existing distribution of access to water sources in the study area. While some own tubewells, others own shallow wells, some others own ponds (individually or as a small group) and some others enjoy greater access to the stream. All of them go about extracting water from these varied sources as though they had nothing to do with one another. One does get the impression that the entire water resource system has been fragmented, with each water source being viewed as an isolated body. This study has illustrated that this is detrimental to both sustainability and equity.

Very often for want of any other alternative, public or government control is recommended to redress the inequities precipitated by private control. In the case of the ponds in the study area for instance, farmers have recommended that the government take control over private ponds in order to ensure greater equity. Government control, however, is not the same as collective control. It often results in control by the bureaucracy, where the collective of water users have little role to play. Neither does government control ensure equitable access as the canals in the study area have illustrated. Even in the case of tubewells, government control has been viewed with caution (Meinzen-Dick 2000).

Defining collective rights or collective control over water requires a good amount of hair splitting. For one it raises questions regarding the scale at which such collective or community institutions are to be institutionalised. Since the 1990s, the emerging literature on Integrated Water Resources Management has emphasised that water as a resource needs to be managed at the level of natural hydrological units such as catchments or river basins, and that institutions for the management of water need to be created at the same level. At the same time, socio political organizations such as the state or the village often claim the right of regulation and distribution of water that is more readily perceived and legally treated as a common good than land (Benda-Beckmann and Benda-Beckmann 2006). Neither the state nor the village is organized along hydrological boundaries. There continues to exist a considerable amount of ambiguity therefore as to who constitutes the collective—a people, a nation state, a village community or a catchment community? (ibid.). In the study area, while the panchayat is growing into an important socio-political institution at the village level, that has both legal and social acceptance, it does not conform to catchment boundaries. Constituting collectives or community institutions at the catchment level implies the imposition of a new institutional entity that can come into conflict with existing ones (Moss 2006). These are issues that required detailed discussions and debates at various levels.

Then again, institutionalising any form of collective ownership or management implies that it confronts the existing system of rights. Collective ownership implies that water needs to be viewed as a common property resource. If one were to envisage such a possibility in the future as far as the study region is concerned, it would imply that all the existing water resources (irrespective of their present private or public status) be turned over to the larger collective or community of water users. Moreover since rights of use to water are derived from land rights, which are unequally distributed, the latter also would be subject to reconsideration. So long as land itself is privately and unequally owned, the scope for achieving a more equitable and sustainable use of water is limited (Bhatia 1992). Land reforms for instance have been considered to be a critical necessity if equitable distribution of irrigation water is to be ensured (Jairath 1985). Introducing collective control and management over land and water would also require changes in the existing legal regime that caters to individual rights and duties, and not to a communal framework of resource management (Ojwang and Juma 1996). Such propositions would be shot down at the first instance by farmers who benefit from the existing status quo. Such proposals would also find little political acceptance. Likely opposition to such ideas, however, does not do away with their relevance in ensuring equity in the long run.

Apart from redressing the iniquitous rights regime, collective control over resources also implies addressing the skewed power relations that exist within communities (Meinzen-Dick and Zwarteveen 2001). In the current situation, the small and marginal farmers who suffer from poor access to water are dependent on the large farmers in many ways for meeting their livelihood needs (see Chapter 6). The possibility of the marginal farmers asserting their rights at a common platform is highly doubtful. Upadhyay therefore argues that while it is important to constitute group rights in water management in order to resolve the existing conflicts surrounding access and control over water, it is equally important to lay down the right of the group members vis a vis one another (Upadhyay 2002).

Collective or community rights to water resources can therefore remove the ills of a private property regime only if it confronts the existing property rights regime that is centred on individual rights. It also requires a confrontation with existing socio political power relations. A reconceptualisation of existing property classifications therefore has to be accompanied by a long drawn out process of social empowerment. The emphasis on participatory processes during the past few decades has led to the constitution and strengthening of various institutional arrangements that attempt to redress inequity in different spheres. In the case of water resources management, watershed associations and water user associations have been put in place to meet the above objective. More recently panchayats too have been constitutionally empowered to address issues related to the management and distribution of water resources at the village level. None of them have, however, confronted the controversial issue of rights to resources (Baumann et al. 2003). As a result, they have not made much of a difference to existing patterns of distribution.

Collective stewardship: Concept for the future?

In the debate over rights, whether individual rights or community rights, the issue of responsibilities, obligations and duties gets sidelined. This is perhaps an outcome of the individual centeredness of legal doctrines. I discuss here the relevance of a few concepts that address the responsibility dimension.

While nation states have been vested with control over natural resources such as water and forests, they have also been viewed as custodians or trustees of these resources. The concept of custodianship or trusteeship is one that emphasizes the responsibility of the state in ensuring the protection of natural resources such as the air, the seas, water and forests, over which private ownership is considered to be 'unjustified' (Iyer 2003). As per this doctrine, the state while being vested with control over critical natural resources, is duty bound to hold the resources of the nation in trust for its citizens, and while doing so ensure that the natural right of any individual or group, or the interests of the public and that of the larger ecosystem are not violated (Singh 1992). These concepts clearly imply that the state is duty bound to prevent the degradation of resources and to prevent the operation of private interests over the use of common property resources.

In the Indian context, the concept of trusteeship has, however, remained at the level of a mere tenet. In practice, the state has rarely appeared as a trustee of natural resources. In the case of water for instance, the state being vested with far ranging proprietary and regulatory powers, is empowered to acquire all water bodies in the name of public interest. While doing so however the state has not been able to ensure the welfare of all members of the 'public'. Neither has it been able to avert the degradation of the resource.

This does not mean that the concept of trusteeship or custodianship is irrelevant. Trusteeship may need to be qualified in order to ensure its stated objectives. Torori et al. opine that certain key issues need to be embodied in the operationalisation of public trust, in order to ensure sustainable and equitable water resources management. These include democratic decision making and equitable distribution of resources, as well as measures that enhance the sustainability of the resource such as enhancement of groundwater recharge and multiple compatible uses of water catchments (Torori et al. 1996).

The relevance of the concept of trusteeship is that it need not be applied to the state alone, but can be applied to other institutions that play a more direct role in the actual management and distribution of water resources, such as the concerned government departments, panchayats, and local level organizations. While this may seem utopian, its relevance in the face of increasingly competitive use of water and increasing extraction of scarce water supplies cannot be undermined.

A concept that bears close resemblance to that of trusteeship is that of stewardship, a concept that is being increasingly emphasized in the context of natural resource protection. It gives emphasis on the need to instil people with a sense of responsibility towards the resources they use. Operationalising such a concept in daily resource use behaviour implies a shift in thinking wherein natural resources are no longer viewed as being at the disposal of human agency. Encouraging stewardship at the level of small communities or user groups implies that community or collective norms evolve with regard to the management and distribution of water resources.

While community management of water resources is assuming increased relevance given the policy emphasis on local level restoration measures, the success of such measures is closely related to the interest and motivation of community members in this regard. People, however, will be encouraged to adopt a more responsible attitude towards the management and use of resources only if they are convinced of the importance of the sustainability dimension. Woodhouse argues that promoting meaningful dialogues between resource users, government officials and locally elected representatives on what constitutes sustainable land and water use would be a step forward towards developing a consensus in this regard and devising regulations towards this end that do not stifle local initiative (Woodhouse 2003). This is a valid suggestion, particularly given the total absence of public debates and discussions on such issues.

Collective stewardship is also relevant in the context of enforcing regulations over land and water use, in order to prevent over-exploitation. Chapter 6 has discussed various restrictions that have been proposed in varying contexts of water use. While state enforcement of such regulations has been the norm, formulation and enforcement of such measures at the community level has been advocated for providing opportunities to instil a sense of responsibility for the conservation of community resources (Bhatia 1992). These issues point to the relevance of facilitating the growth of community level institutions that assume the responsibility for ensuring a sustainable and equitable mode of water resources management.

The present effort at democratic decentralisation in the state has enhanced the possibilities for such local level deliberations. In the case of water, the panchayat can take an initiative to generate baseline data on water resources which can serve as an important starting point to discussions on sustainable and equitable water resources management. The panchayat can also facilitate dialogues on this issue among the public and foster a collective understanding of the limits to the conventional pattern of water resource development. Such a collective consensus is important in ensuring sustainable land and water use patterns. This does not undermine the importance of other facilitating conditions such as state directives and policies that emphasize on sustainability as well as legal doctrines and regulations in this regard. If the current thrust on decentralisation is to have any positive impact on sustainable land and water management, fostering a spirit of collective stewardship is important. In a nutshell, addressing the water crisis that many parts of Palakkad district face today necessitates reconsiderations and re-evaluations on many fronts. In the long term, existing irrigation and agricultural policies need to be reoriented to the requirements of sustainable water resources management. Existing irrigation technologies, viz. the reservoir based canal systems, and the growing spread of the lift irrigation and tubewell technologies also need to be evaluated from this angle. An important question that needs to be asked is for how long can the existing irrigation policies and technologies support the water intensive cropping patterns that they presently encourage? In the short term, functioning of the existing canal irrigation infrastructure and distribution systems needs to be improved. However problems of inequitable distribution will not be resolved through repairs and maintenance and through the constitution of local water user associations alone. Conditions need to evolve whereby water allocation and distribution rules are followed by the large and the small farmers alike.

A re-envisioning of property rights over land and water such that both sustainability and equity is ensured is another challenge that is thrown up by the present water crisis. This thesis has emphasized the relevance of conceptualising property rights in ecological terms. It has also emphasised the need to restrict private encroachments into commonly owned water resources. The concept of a 'Water Reform' is both timely and relevant. However, unlike the land reform initiative that was focused on the issue of distribution alone, an agenda for Water Reform should address the way in which the resource is being managed. It should also revisit the concepts of property in both land and water that inform the current pattern of distribution. At present this appears to be impossible, even 'unthinkable'. It, however, remains inevitable to any long lasting reform process.

References

Abeyratne, S. D. 1990. Rice, Rehabilitation and Rural Change: Social Organization and State Intervention in Small-Scale Irrigation Systems in Sri Lanka. Ph.D. Thesis, Cornell University.

Amruth, M. 2004. Forest-Agriculture Linkage and its Implications on Forest Management: A Study of Delampady *panchayat,* Kasargod district, Kerala. Discussion Paper No. 69, Kerala Research Programme on Local Level Development. Trivandrum: Center for Development Studies.

Anderson, Terry L. and P.J. Hill. 1977. From Free Grass to Fences: Transforming the Commons of the American West. In *Managing the Commons*, eds. Garrett Hardin and John Baden, pp. 200–216.. New York: W.H. Freeman And Company.

Athreya, Venkatesh B., Goran Djurfeldt and Staffan Lindberg. 1990. *Barriers Broken. Production Relations and Agrarian Change in Tamil Nadu.* New Delhi: Sage Publications.

Balakrishnan, Pulapre. 1999. Land Reforms and the Question of Food in Kerala, *Economic and Political Weekly*, March 22.

Bandyopadhyay, Jayanta. 1987. Political Ecology of Drought and Water Scarcity. Need for an Ecological Water Resources Policy. *Economic and Political Weekly*, December 12, 1987: 2159–2169.

Bandyopadhyay, Jayanta. 2006. Criteria for a Holistic Framework for Water Systems Management in India. In *Integrated Water Resources Management. Global Theory, Emerging Practice and Local Needs*, eds. Peter P. Mollinga, Ajaya Dixit and Kusum Athukorala, pp. 145–171. New Delhi: Sage Publications.

Barah, B.C., D. N. Reddy and T. Sudhakar. 1996. Decline of Traditional Water Harvesting Systems. In *Traditional Water Harvesting Systems. (An Ecological Economic Appraisal)*, ed. B. C. Barah, pp. 149–162. New Delhi: New Age International Limited.

Barker, Randolph and Francois Molle. 2004. *Evolution of Irrigation in South and Southeast Asia.* Comprehensive Assessment Research Report 5. Colombo, Sri Lanka: IWMI.

Baumann, P; R Ramakrishnan; M Dubey; R K Raman and J Farrington. 2003. *Institutional Alternatives and Options for Decentralised Natural Resource Management in India.* ODI Working Paper No. 230. UK: Overseas Development Institute.

Benda Beckmann, F von. 2001. Between Free Riders and Free Raiders: Property Rights and Soil Degradation in Context. In *Economic Policy and Sustainable Land Use. Recent Advances in Quantitative Analysis for Developing Countries,* eds, Heerink, Nico, Herman van Keulen and Marijke Kuiper, Heidelberg, New York: Physica-Verlag.

Benda-Beckmann F. von and K. von Benda-Beckmann. 2003. Water, Human Rights And Legal Pluralism. *Water Nepal,* 9/10(1/2): 63–76.

Benda-Beckmann, F. von and K. von Benda-Beckmann. 2006. Gender, Water Rights and Irrigation in Nepal. In *Fluid Bonds. Views on Gender and Water,* ed. Kuntala Lahiri-Dutt, Kolkata: STREE.

Benda-Beckmann, F. von and K. von Benda-Beckmann. 2006. How Communal is Communal and Whose Communal is It? Lessons from Minangkabau. In *Changing Properties of Property,* eds. Benda-Beckmann F. von, K. von Benda-Beckmann and Melanie Wiber, New York: Berghahn Books.

Benda-Beckmann, F. von, and K. von Benda-Beckmann. 1999. Community Based Tenurial Rights. Emancipation or Indirect Rule? In *Papers of the XIth International Congress "Folk Law and Legal Pluralism: Socieites in Transformation"* of the Commission on Folk Law and Legal Pluralism, Moscow, August 18–22, 1997, eds. K. von Benda-Beckmann and H.F. Finkler.

Benda-Beckmann, F. von, K. von Benda-Beckmann and H.L.J. Spiertz. 1996. Water Rights and Policy. In *The Role of Law in Natural Resource Management,* eds. H.L.J. Spiertz and Melanie Wiber, The Hague: VUGA.

Benda-Beckmann, F. von. 1992. Uncommon Questions About Property Rights. *Recht Der Werkelijkheid,* 1992/1. Gravenhage: VUGA.

Benda-Beckmann, F. von; K. von Benda-Beckmann; Melanie Wiber. 2003. The Properties of Property. In *Changing Properties of Property,* eds. F. von Benda-Beckmann, K. von Benda-Beckmann and Melanie Wiber, pp. 1–39. New York: Berghahn Books.

Bennett, J.W. 1993. Human ecology as human behaviour: essays in environment and development anthropology.

Berkes, Fikret, ed. 1989. Common Property Resources: Ecology and Community-Based Sustainable Development. London: Belhaven.

Berkes, Fikret. 1996. Social Systems, Ecological Systems, and Property Rights. In *Rights to Nature. Ecological, Economic, Cultural, and Political Principles of Institutions for the Environment*, eds. Susan Hanna, Carl Folke and Karl-Goran Maler, pp. 87–110. Washington D.C.: Island Press.

Bhargava, Meena and John F. 2002. Defining Property Rights in Land in Colonial India: Gorakhpur Region in the Indo-Gangetic Plain. In In *Land, Property and the Environment*, ed. John F. Richards, 235–262. Oakland, California: Institute for Contemporary Studies.

Bharwada, Charul and Vinay Mahajan. 2002. Drinking Water Crisis in Kutch: A Natural Phenomenon?. *Economic and Political Weekly*, November 30, 2002.

Bhatia, Bela. 1992. Lush Fields and Parched Throats. Political Economy of Groundwater in Gujarat. *Economic and Political Weekly*, December 19–26, 1992.

Bhattathiripad, T N N. 2003. Water Conservation. In *Managing Water and Water Users. Experiences from Kerala*, eds. Sooryamoothy, R and Antony Palackal S J, pp. 79–89. Maryland: University Press of America.

Bhullar, A S and R S Sidhu. 2006. Integrated Land and Water Use. A Case Study of Punjab. *Economic and Political Weekly*, December 30, 2006: 5353–5357.

Bijoy, C. R. 2006. Kerala's Plachimada Struggle. A Narrative on Water and Governance Rights. *Economic and Political Weekly*, Vol XLI No. 41: 4332–4339.

Bolding, Alex. 2004. In Hot Water. A study on sociotechnical intervention models and practices of water use in smallholder agriculture, Nyanyadzi catchment, Zimbabwe. Ph.D. dissertation, Wageningen University, Wageningen, the Netherlands.

Bottrall, A.F. 1992. Fits and misfits over time and space: technologies and institutions of water development for South Asian agriculture. *Contemporary South Asia*, 1(2): 227–247.

Bromley, Daniel W. 1989. Property Relations and Economic Development: The Other Land Reform. *World Development*, 17(6): 867–877.

Bromley, Daniel W. 1991. Environment and Economy. Property Rights and Public Policy. Cambridge: Blackwell.

Bromley, Daniel W. 2002. Property Regimes for Sustainable Resource Management. In *Land, Property and the Environment*, ed. John F. Richards, 338–354. Oakland, California: Institute for Contemporary Studies.

Bromley, Daniel W. and Michael M. Cernea. 1989. *The Management of Common Property Natural Resources. Some Conceptual and Operational Fallacies*. World Bank Discussion Paper No. 57, Washington, D.C.: The World Bank.

Bruijnzeel, L.A., 1990. Hydrology of moist forests and the effects of conversion: a state of knowledge review. In *UNESCO-Int. Inst. for Aerospace Survey and Earth Science*, Publication of the Humid Tropics Programme. Amsterdam: Free University.

Buchanan, Francis. 1870. *A Journey from Madras through the Countries of Mysore, Canara, and Malabar*, Vol. I and II. Madras: Higginbotham's and Company.

Burke, Jacob J. 2000. Land and water systems: managing the hydrological risk. *Natural Resources Forum*. 24(2): 123–136.

Burke, Jacob J., Claude Sauveplane and Marcus Moench. 1999. Groundwater management and socio-economic responses. *Natural Resources Forum*, 23(1999): 303–313.

Calder, I.R. 1998. *Water-Resource and Land-Use Issues*. SWIM Paper 3, Colombo, Sri Lanka: IWMI.

Campbell, Bruce M.S. and Ricardo A. Godoy. 1986. Commonfield Agriculture: The Andes and Medieval England Compared. In *Proceedings of the Conference on Common Property Resource Management*, National Academy of Sciences, pp. 323–358. Washington: National Academy Press.

Census of India, 1961. District Census Handbook Palghat.

Census of India, 1971. District Census Handbook Palghat.

Census of India, 1991. District Census Handbook Palghat.

Census of India, 2001. District Census Handbook, Palghat.

Centre for Water Resources Development and Management (CWRDM). 1981. *Seminar papers on Resource Potential of Kerala*, Volume I, Calicut: CWRDM.

Centre for Water Resources Development and Management (CWRDM). n.d. Water Scarcity in Kerala- Policy Guidelines for Better Management, Kozhikode: CWRDM.

Chambers, Robert. 1977. Men and Water: the Organisation and Operation of Irrigation. In *Green Revolution? Technology and Change in Rice-Growing Areas of Tamil Nadu and Sri Lanka*, ed. B.H. Farmer, pp. 340–363. London and Basingstoke: Macmillan.

Chopra, Kanchan; G.K. Kadekodi and M.N. Murty. 1990. *Participatory Development: People and Common Property Resources*, New Delhi: Sage Publications.

Christodoulou, Demetrios. 1990. The Unpromised Land. Agrarian Reform and Conflict Worldwide. London: Zed Books.

Chundamannail. M 1993. *History of Forest Management in Kerala.* KFRI Research Report No. 89, Peechi: Kerala Forest Research Institute.

Committee on Plan Projects, 1960. *Report on Minor Irrigation Works (Kerala State).* Committee on Plan Projects (Irrigation and Power Team), New Delhi: Government of India.

Cotula, Lorenzo. 2006. Land and water rights in the Sahel. Tenure challenges of improving access to water for agriculture. IIED Issue Paper No. 139. London: IIED.

Crow, Ben and Farhana Sultana. 2002. Gender, Class, and Access to Water: Three Cases in a Poor and Crowded Delta. *Society and Natural Resources*, 15(8): 709–724.

D'Souza, Rohan. 2003. Supply-Side Hydrology in India. The Last Gasp. *Economic and Political Weekly*. September 6, 2003.

Demsetz, Harold. 1967. Toward a Theory of Property Rights. *American Economic Review*, 62: 347–59.

Dijk, Han van. 1996. Land Tenure, Territoriality, and Ecological Instability: A Sahelian Case Study. In *The Role of Law in Natural Resource Management,* eds. H.L.J. Spiertz and Melaine Wiber, The Hague: VUGA.

District Planning Comittee, n.d. *Zilla Vikasana Padhati (In Malayalam).* (District Development Plan)-Part 1 (a). Palakkad District Planning Committee.

Dubash, Navroz K. 2007. The Local Politics of Groundwater in North Gujarat. In *Waterscapes. The Cultural Politics of a Natural Resource.* Baviskar, Amita (ed.), Uttaranchal: Permanent Black.

Elavenchery Panchayat, 1996. Vikasana Rekha (In Malayalam).

Falkenmark, M. and C. Widstrand. 1992. *Population and water resources: a delicate balance.* Population Bulletin, Population Reference Bureau.

Falkenmark, M. and J. Lundqvist. 1998. Towards water security: Political determination and human adaptation crucial. *Natural Resources Forum*, 21(1): 37–51.

Frankel, Francine R. 1972. *India's Green Revolution. Economic Gains and Political Costs.* Princeton, New Jersey: Princeton University Press.

Freyfogle, E. T. 1996. Ethics, Community, and Private Land, *Ecology Law Quarterly,* 23: 631–61.

Freyfogle, E. T. 1999. The Particulars of Owning. *Ecology Law Quarterly*, 25:574

Ganesh, K.N. 1991. Ownership and Control over Land in Medieval Kerala: Janmam-kanam relations during the 16th–18th centuries. *Indian Economic and Social History Review*, 28 93 0: 299–321.

Gatzweiler, F.W. and K. Hagedorn. 2002. The evolution of institutions in transition. *International Journal of Agricultural Resources Governance and Ecology*, 2(1): 37–58.

Geertz, Clifford. 1963. Agricultural Involution: The Processes of Ecological Change in Indonesia. Berkeley: University of California Press.

George, Abey and Jyothi Krishnan. 2000. An Overview of the Electricity Sector in Kerala. In *The Reform Process and Regulatory Commissions in the Electricity Sector: Developments in Different States of India*', Prayas and Focus on the Global South.

George, Alex. 1987. Social and Economic Aspects of Attached Labourers in Kuttanad Agriculture. *Economic and Political Weekly*, December 26, 1987, pp. A–141 to A–150

George, M.V. and N.G. Nair. 1982. Irrigation and Agricultural Development in Kerala. In *Agricultural Development in Kerala*, ed. P.P. Pillai. New Delhi: Agricole Publishing Academy.

George, P. S. and Chandan Mukherjee. 1986. *Rice Economy of Kerala: A disaggregate analysis of Performance*. Working Paper No. 213, Trivandrum: Centre for Development Studies.

Gilmartin, David. 2003. Water and Waste: Nature, Productivity and Colonialism in the Indus Basin. *Economic and Political Weekly*, Nov 29, 2003.

Gleick, Peter H. (ed.), 1993. Water in Crisis: A Guide to the World's Fresh Water Resources, Oxford University Press.

Glennon, Robert. 2002. Water Follies. Groundwater Pumping and the Fate of America's Fresh Waters. Washington: Island Press.

Global Water Partnership, 2000. *Integrated Water Resources Management*. TAC Background Paper 4, Global Water Partnership Secretariat, Stockholm.

Goldsmith, Edward and Nicholas Hildyard. 1984. *The Social and Environmental Effects of Large Dams: Vol I*. Cornwall, UK: Wadebridge Ecological Centre.

Gopikuttan, G. and K.N. Parameswara Kurup. 2004. Paddy Land Conversion in Kerala. An inquiry into ecological and economic aspects in a midland watershed region. Kerala Research Programme on Local Level Development. Trivandrum: Centre for Development Studies.

Government of Kerala, 1976. *Meteorological Factors and Crop Pattern in Kerala*. Trivandrum: Bureau of Economics and Statistics.

Government of Kerala, 1977. *Economic Review* 1976. Trivandrum: State Planning Board.

Government of Kerala, 1981a. *Agricultural Census 1976–77. Report for Kerala State, Vol I.* Trivandrum: Directorate of Economics and Statistics.

Government of Kerala, 1981b. *Command Area Development Programme In Kerala. A Brief Review*. Review Paper no. 1/81, Planning and Economic Affairs Department. Trivandrum: Govt. of Kerala.

Government of Kerala, 1998. *Revamping and Consolidation Programme. Gayatri Project Stage I and II.* Project Report, October 1998. Irrigation Division, Chittur-Palakkad.

Government of Kerala, 2001a. *Panchayat Level Statistics 2001*. Trivandrum: Department of Economics and Statistics.

Government of Kerala, 2001b. *Statistics for Planning 2001.* Trivandrum: Department of Economics and Statistics.

Government of Kerala, 1994. *Economic Review* 1993. Trivandrum: State Planning Board.

Government of Kerala, 1995. *Economic Review* 1994. Trivandrum: State Planning Board.

Government of Kerala, 1996. *Economic Review* 1995. Trivandrum: State Planning Board.

Government of Kerala, 1997. *Economic Review* 1996. Trivandrum: State Planning Board.

Government of Kerala, 1998. *Economic Review* 1997. Trivandrum: State Planning Board.

Government of Kerala, 1999. *Economic Review 1998.* Trivandrum: State Planning Board.

Government of Kerala, 2002a. *Economic Review 2001*. Trivandrum: State Planning Board.

Government of Kerala, 2002b. *Ongoing Major and Medium Irrigation Projects in Kerala. A Quick Study*. Evaluation Division, State Planning Board, Thiruvananthapuram.

Government of Kerala, 2003. *Economic Review 2002.* Trivandrum: State Planning Board.

Government of Kerala, 2003a. District Handbooks of Kerala. PALAKKAD. Trivandrum: Department of Information and Public Relations.

Government of Kerala, 2004. *Economic Review 2003.* Trivandrum: State Planning Board.

Government of Kerala, 2005. *Economic Review 2004.* Trivandrum: State Planning Board.

Government of Kerala, 2006a. *Agricultural Statistics 2005–06.* Thiruvananthapuram: Department of Economics and Statistics.

Government of Kerala, 2006b. *Paddy Cultivation in Kerala (1995–96 to 2004–05).* Thiruvananthapuram: Department of Economics and Statistics.

Government of Kerala, 2006c. *Panchayat Level Statistics 2006.* Trivandrum: Department of Economics and Statistics.

Government of Kerala, 2006d. *Economic Review 2005.* Trivandrum: State Planning Board.

Government of Kerala, 2007a. *District Level Database 2006-Palakkad.* Government of Kerala: Department of Local Self-Government, and KILA.

Government of Kerala, 2007b. *Economic Review 2006.* Trivandrum: State Planning Board.

Government of Kerala, 2007c. Guidelines for the Preparation of Annual Plan (2007–08) and Eleventh Five-Year Plan (2007–12). Trivandrum: Department of Local Self-Government.

Government of Kerala, n. d. National Water Management Project. Project Report on Chitturpuzha Scheme. Irrigation Division, Chittur-Palakkad.

Government Rice Commission, 1999. *Report of the Expert Committee on Paddy Cultivation.* Trivandrum: Government of Kerala.

Gunneli, Yanni and Anupama Krishnamurthy. 2003. Past and Present Status of Runoff Harvesting Systems in Dryland Peninsular India: A Critical Review. *Ambio*, 32(4): 320–324.

Hanks, L.M. 1972. Rice and Man: Agricultural Ecology in Southeast Asia.

Hann, C.M. 1998. Introduction: the embeddedness of property. In *Property Relations. Renewing the anthropological tradition*, ed. C.M. Hann. Cambridge: Cambridge University Press.

Hanna, Susan; Carl Folke; Karl-Goran Maler. 1996. Property Rights and the Natural Environment. In *Rights to Nature. Ecological, Economic, Cultural, and Political Principles of Institutions for the Environment*, eds. Susan Hanna, Carl Folke and Karl-Goran Maler, pp. 1–12. Washington D.C.: Island Press.

Hardiman, David. 2007. The Politics of Water Scarcity in Gujarat. In *Waterscapes. The Cultural Politics of a Natural Resource*. Baviskar, Amita (ed.), Uttaranchal: Permanent Black.

Hardin, Garrett. 1968. The Tragedy of the Commons. *Science*, 1968, pp. 1243–1248.

Hart, Henry C. 1978. Anarchy, Paternalism, or Collective Responsibility Under the Canals? *Economic and Political Weekly*, December 1978.

Herring, Ronald J. 1983. *Land to the Tiller*. New Haven: Yale University Press.

Herring, Ronald J. 1988. *The Commons and its 'Tragedy' as Analytical Framework: Understanding Environmental Degradation*. Department of Political Science, Northwestern University, Evanston, IL.

Herring, Ronald J. 1990. Rethinking The Commons. *Agriculture and Human Values*, Spring 1990: 88–104.

Herring, Ronald J. 2000. *Political Conditions For Agrarian Reform And Poverty Alleviation*. IDS Discussion Paper no. 375. Brighton, UK: IDS, University of Sussex.

Herring, Ronald J. 2002. State Property Rights in Nature (with Special Reference to India). In *Land, Property and the Environment*, ed. John F. Richards, 263–297. Oakland, California: Institute for Contemporary Studies.

Hodgson, S. 2004. Land and water-the rights interface. Rome: FAO.

Hufschmidt, M. M and K G Tejwani. 1993. Integrated Water Resource Management: Meeting the Sustainability Challenge. France: UNESCO.

Huggins, Christopher. 2000. Rural Water Tenure in East Africa. A Comparative Study of Legal Regimes and Community Responses to Changing Tenure Patterns in Tanzania and Kenya. BASIS Brief, African Centre for Technology Studies, Nairobi, Kenya.

Innes, C.A. 1908. *Malabar Gazetteer*. Trivandrum: Kerala Gazetteers Department, Government Of Kerala (Second Reprint 1997).

Iyengar, Sudarshan, Nimisha Shukla. 2002. Common Property Land Resources in India: Some Issues in Regeneration and Management. In *Institutionalizing Common Pool Resources*, ed. Dinesh K. Marothia. New Delhi: Concept Publishing Company.

Iyer, Ramaswamy R. 1998. Water Resource Planning. Changing Perspecties. *Economic and Political Weekly*, December 12, 1998: 3198–3205.

Iyer, Ramaswamy R. 2003. *Water. Perspectives, Issues, Concerns*. New Delhi: Sage Publications.

Iyer, Ramaswamy. 2001. Water: Charting a Course for the Future-II. *Economic and Political Weekly*. April 14–20, 2001.

Jairath, Jasveen. 1985. Technical and Institutional Factors in Utilisation of Irrigation. A Case Study of Public Canals in Punjab. *Economic and Political Weekly*, 20(13).

Janakarajan, 2002. Wells and Illfare: An Overview of Groundwater Use and Abuse in Tamil Nadu, South India. Paper Presented at the IWMI-TATA Annual Partners Meet, 2002.

Jannuzi, Tomasson F. 1994. India's Persistent Dilemma. The Political Economy of Agrarian Reform. Boulder: Westview Press.

Jayaraman, T. K. 1980. A Case for Professionalisation of Water Management in Irrigation Projects in India. In In *Proceedings of Seminar on Water Management Practices in Kerala*. Calicut, 11–12, October 1980. CWRDM, Calicut.

Jha, C.S., C. B. S. Dutt and K. S. Bawa. 2000. Deforestation and land use changes in Western Ghats, India. *Current Science*, 79(2): 231–238).

Jodha, N. S. 1986. Common Property Resources and Rural Poor in Dry Regions of India. *Economic and Political Weekly*, 21, 1169–81, July 5.

Jose, Sunny. 1991. Group Farming in Kerala. An Illustrative Study. Mphil Thesis, Centre for Development Studies, Trivandrum.

Joseph, C. J. 2001. *Beneficiary Participation in Irrigation Water Management: The Kerala Experience*. Discussion Paper No. 36, Kerala Research Programme on Local Level Development, Trivandrum: Centre for Development Studies.

Jurriens, M.; Peter P. Mollinga; P. Wester. 1996. *Scarcity By Design. Protective Irrigation in India and Pakistan*. Liquid Gold Paper 1, Wageningen: Wageningen Agricultural University.

Kannan, K. P. and K. Pushpangadan. 1988. *Agricultural Stagnation and Economic Growth in Kerala. An Exploratory Analysis.* Working Paper No. 227, Trivandrum: Centre for Development Studies.

Kannan, K. P. and K. Pushpangadan. 1990. *Dissecting Agricultural Stagnation in Kerala: An Analysis Across Crops, Seasons and Regions.* Working Paper No. 238, Trivandrum: Centre for Development Studies.

Kareem, C.K. 1976. *Palghat*. Kerala District Gazetteers. Trivandrum: Govt. Of Kerala.

Kendy, Eloise, David J. Molden, Tammo S. Steenhuis, Changming Liu and Jinxia Wang. 2003. *Policies Drain the North China Plain. Agricultural Policy and Groundwater Depletion in Luancheng County, 1949–2000*. Research Report 71, Colombo, Sri Lanka: International Water Management Institute.

Kerala Panchayati Raj Act, 1994.

Kerala State Land Use Board, 2001. *Vibhava Bhoopada Nirmana Report (In Malayalam)*, Report on Resource Map Preparation, Kollengode and Elavenchery Grama Panchayats, Palakkad Zilla.

Kerala Statistical Institute, 1994. *Conversion of Paddy Land Kerala.* Trivandrum: Kerala Statistical Institute.

Klug, Heinz. 2002. Straining the Law: Conflicting Legal Premises and the Governance of Aquatic Resources. *Society and Natural Resources,* 15: 693–707.

Kumar, B. M. 2005. Land use in Kerala: changing scenarios and shifting paradigms. *Journal of Tropical Agriculture,* 42(1–2): 1–12.

Kurnia, Ganjar; Teten W. Avianto; Bryan Randolph Bruns. 2000. Farmers, Factories and the Dynamics of Water Allocation in West Java. In *Negotiating Water Rights.* Bruns, B.R. and R. S. Meinzen-Dick (ed.) pp. 292–314, IFPRI., New Delhi: Vistaar Publications, London: Intermediate Technology Press.

Kuruvilla, P. K. 1967. Irrigated Agriculture in Kerala. Minor Irrigation in Kerala. In *Irrigated Agriculture in Kerala and New Perspectivs for Building Programme in Kerala.* Proceedings of the Symposium held at Peechi, May 30–31, 1967. Kerala Engineering Research Institute, Peechi. GOK PWD.

Lam, Wai F. 1998. Governing Irrigation Systems in Nepal: Institutions, Infrastructure, and Collective Action. Oakland, CA: ICS Press.

Logan, William. 1887. *Malabar* Vol I. Madras: Government Press, Reprint 1951.

McCay, Bonnie J. and James M. Acheson. 1987. Human Ecology of the Commons. In *The Question of the Commons. The Culture and Ecology of Communal Resources,* eds. Bonnie J. McCay and James M. Acheson, pp. 1–36. Tucson: The University of Arizona Press.

McKean, Margaret A. 1996. Management of Traditional Common Lands (Iriachi) in Japan. In *Proceedings of the Conference on Common Property Resource Management,* Washington D.C.: National Research Council.

McKean, Margaret A. 2000. Common Property: What Is It, What Is It Good for, and What Makes It Work? In *People and Forests. Communities, Institutions, and Governance.* Gibson, C., M. McKean and E. Ostrom (ed.), Cambridge: MIT Press.

Mehta, Lyla. 2000. *Water For The Twenty-First Century: Challenges And Misconceptions.* IDS Working Paper No. 111, Brighton, UK: IDS, University of Sussex.

Mehta, Lyla. 2003. Contexts and Constructions of Water Scarcity. *Economic and Political Weekly* November 29, 2003.

Mehta, Lyla. 2007. Whose scarcity? Whose property? The case of water in western India. *Land Use Policy,* 24(2007): 654–663.

Meinzen-Dick, R. S. and B. R. Bruns. 2000. Negotiating Water Rights: Introduction. In *Negotiating Water Rights*. Bruns, B.R. and R. S. Meinzen-Dick (ed.) pp. 23–41. IFPRI, New Delhi: Vistaar Publications, London: Intermediate Technology Press.

Meinzen-Dick, R.. and Margreet Zwarteveen. 2001. Gender Dimensions of Community Resource Management. The Case of Water Users' Associations in South Asia. In *Communities and the Environment. Ethnicity, Gender and the State in Community-Based Conservation*. Agrawal, Arun and Clark C. Gibson (ed.) Rutgers University Press.

Meinzen-Dick, R.S. and Rajendra Pradhan. 2001. Implications of legal pluralism for natural resource management. *IDS Bulletin,* 32(4).

Meinzen-Dick, Ruth S. 2000. Public, Private, and Shared Waters: Groundwater Markets and Access in Pakistan. In *Negotiating Water Rights*. Bruns, B.R. and R. S. Meinzen-Dick (ed.) pp. 245–263, IFPRI., New Delhi: Vistaar Publications, London: Intermediate Technology Press.

Mencher, Joan P. 1980. The Lessons and Non-Lessons of Kerala. Agricultural Labourers and Poverty. *Economic and Political Weekly*, October 1980.

Mencher, Joan P. 1998. Ecology and Social Structure: A Comparative Analysis. In *Social Ecology*, ed. Ramachandra Guha, pp. 42–81. Delhi: Oxford University Press.

Mendis, D.L.O. 1999. Hydraulic Engineering Versus Water And Soil Conservation Ecosystems: Lessons From The History Of The Rise And Fall Of Sri Lanka's Ancient Irrigation Systems. *Water Nepal*, 7(1): 49–89.

Menon, Dilip M. 1994. Caste, Nationalism and Communism in Malabar (1900–1948). Cambridge: Cambridge University Press.

Menon, Madhava T. 1983. *Rice in Kerala-Prospects and Strategies*. Paper Presented at Seminar on Stagnation of Rice Production in Kerala. Kerala Agricultural University and Kerala State Planning Board, 1st–3rd July, 1983, College of Agriculture, Vellayani.

Mitchell, Bruce. 2005. Integrated water resource management, institutional arrangements, and land-use planning. *Environment and Planning A*, 37: 1335–1352.

Molden, D C de Fraiture. 2004. I*nvesting in Water for Food, Ecosystems and Livelihoods.* Blue Paper, Comprehensive Assessment of Water Management. Colombo, Sri Lanka: IWMI.

Molle, Francois and Peter Mollinga. 2003. Water poverty indicators: conceptual problems and policy issues. *Water Policy* 5(2003): 529–544.

Molle, Francois, Tushaar Shah and Randy Barker. 2003. *The Groundswell of Pumps: Multilevel Impacts of a Silent Revolution.* Paper Prepared for the ICID-Asia Meeting, Taiwan, November, 2003.

Molle, Francois. 2003. *The 'closure' of river basins: trajectories and societal responses.* Paper prepared for the 3rd Conference of the International Water History Association, Alexandra, Egypt, December 11–14, 2003.

Molle, Francois. 2005. *Elements for a political ecology of river basin development: The case of the Chao Phraya river basin, Thailand.* Paper Presented to the 4th Conference of the International Water History Association, December 2005, Paris.

Mollinga, P. P. 2003. On The Waterfront. Water Distribution, Technology and Agrarian Change in a South Indian Canal Irrigation System. Wageningen University Water Series. Hyderabad: Orient Longman.

Mollinga, Peter P. 1998. On The Waterfront. Water Distribution, Technology and Agrarian Change in a South Indian Canal Irrigation System. Ph.D. Diss., Wageningen Agricultural University.

Mollinga, Peter P. 2000. The Inevitability of Reform: Towards Alternative Approaches for Canal Irrigation Development in India. In *Participatory Irrigation Management-Paradigm for the Twenty First Century*, eds. Joshi, L. K. and Rakesh Hooja, Vol I, Jaipur: Rawat Publications.

Mollinga, Peter P. 2006. IWRM in South Asia: A Concept Looking for a Constituency. In *Integrated Water Resources Management. Global Theory, Emerging Practice and Local Needs*, eds. Peter P. Mollinga, Ajaya Dixit and Kusum Athukorala, pp. 21–37. New Delhi: Sage Publications.

Moss, Timothy. 2006. Solving Problems of 'Fit' at the Expense of Problems of 'Interplay'? The Spatial Reorganisation of Water Management following the EU Water Framework Directive. In *Integrated Water Resources Management. Global Theory, Emerging Practice and Local Needs*, eds. Peter P. Mollinga, Ajaya Dixit and Kusum Athukorala, pp. 64–108. New Delhi: Sage Publications.

Mukhopadhyay, Pranab 2005. Now that your land is my land…does it matter? A case study in Western India. *Environment and Development Economics*, 10: 87–96.

Mukundan, T. M. 1996. The *ERY* Systems of South India. In *Traditional Water Harvesting Systems. (An Ecological Economic Appraisal)*, ed. B. C. Barah, pp. 71–100. New Delhi: New Age International Limited.

Nadkarni, M.V. 1987. Agricultural Development And Ecology-An Economist's View. *Indian Journal of Agricultural Economics*, 42(3): 359–375.

Nair, Janardhanan Nair. 1981. Report of The One Man Commission On The Problems of Paddy Cultivators in Kerala 1981. One Man Commission.

Nair, K N and Vineetha Menon. 2006. Lease Farming in Kerala. Findings from Micro Level Studies. *Economic and Political Weekly*, Vol XLI No 26.

Nair, Sathis Chandran. 1991. *The Southern Western Ghats: a biodiversity conservation plan*. Studies in Ecology and Sustainable Development- 4. New Delhi: INTACH.

Nair, Sathis Chandran. 2004. A Note on the Ecological Impact of the Proposed Pathrakadavu Hydro-Electric Project (PHEP) in the Kunthi River, Palakkad District of Kerala. Downloadable from http:// www. salimalifoundation.org/ pathrakkadavu.html

Nair, Shyamasundaran K.N. 1999. Resource Management For Sustainable Development of Keralam's Agriculture. In *Rethinking Development: Kerala's Development Experience*, ed. M.A. Oommen, pp. 269–278. New Delhi: Concept Publishing Company.

Nalapat, M. D. 1978. Irrigation Muddle. *Economic and Political Weekly*, May 20, 1978.

Narayana D, V C V Ratnam and K Narayanan Nair. 1982. An Approach to Study of Irrigation. Case of Kanyakumari District. *Economic and Political Weekly*, September 1982: A-85- A-102.

Narayana, D. and K. Narayanan Nair. 1983. *Linking Irrigation with Development. The Kerala Experience.* Working Paper No. 161, Trivandrum: Centre for Development Studies.

Narayanan, N.C. 2002. Technical Fixes and Social Relations. The Case of Group Farming in Kerala, India. Working Paper Series No. 367. The Hague: Institute of Social Studies.

Netting, Robert McC. 1974. Agrarian Ecology. *Annual Review of Anthropology*, Vol 3, 1974: 21–56.

Netting, Robert McC. 1993. Smallholders, Householders. Farm Families and the Ecology of Intensive, Sustainable Agriculture. Stanford, California: Stanford University Press.

Ojwang, J.B. and Calestous Juma. 1996. Towards Ecological Jurisprudence. In *In Land We Trust. Environment, Private Property and Constitutional Change*, eds. Juma, Calestous and J. B. Ojwang, Nairobi: Initiatives Publishers.

Oommen, M.A. 1971. Land Reforms and Socio-Economic Change in Kerala. An Introductory Study. Madras: The Christian Literature Society.

Orindi, Victor and Chris Huggins. 2005. The dynamic relationship between property rights, water resource management and poverty in Lake Victoria Basin. Paper presented at the International Workshop on '*African Water Laws: Plural Legislative Frameworks for Rural Water management in Africa*', 26–28 January 2005, Johannesburg, South Africa.

Ostrom, Elinor. 1990. Governing the Commons: The Evolution of Institutions for Collective Action, Cambridge: Cambridge University Press.

Ostrom, Elinor. 1992. Crafting Institutions For Self-Governing Irrigation Systems. San Francisco: ICS Press. Bharwada, Charul and Vinay Mahajan 2002 Drinking Water Crisis in Kutch: A Natural Phenomenon? EPW, November 30, 2002.

Pandian, M. S. S. 1990. The Political Economy of Agrarian Change: Nanchilnadu 1880–1939. New Delhi: Sage Publications.

Pant, Niranjan. 1981. Utilisation of Canal Water below Outlet in Kosi Irrigation Project. Administrative and Community-Level Solutions, *Economic and Political Weekly*, September 1981.

Parasuraman, S and Sohini Sengupta 2001. World Commission on Dams. Democratic Means for Sustainable Ends. *Economic and Political Weekly* May 26, 2001.

Pereira, H.C. 1973. Land Use and Water Resources in Temperate and Tropical Climates. London: Cambridge University Press.

Peters, Pauline E. 1984. 'Struggles over Water, Struggles over Meaning: Cattle, Water and the State in Botswana'. *Africa*, 54 (3): 29–49.

Pillai, P.P. ed. 1982. Agricultural Development in Kerala. New Delhi: Agricole Publishing Academy.

Postel, Sandra. 1984. Water: Rethinking Management in an Age of Scarcity. Worldwatch Paper 62. Washington D.C.: Worldwatch Institute.

Postel, Sandra. 1999. Pillar of Sand. Can The Irrigation Miracle Last? Washington D.C.: Worldwatch Institute.

Prabhakaran, 2003. Nila: A Dying River with a Cultural Heritage. In *Managing Water and Water Users. Experiences from Kerala,* eds. Sooryamoothy, R and Antony Palackal S J, pp. 57–66. Maryland: University Press of America.

Prabhakaran, 2004. 'Nila dried up owing to water diversions: experts', *The Hindu*, Saturday, November 6th, 2004

Prakash, B.A. 1987. Agricultural Development of Kerala from 1800 AD to 1980 AD: A Survey of Studies. Working Paper No. 220, Centre for Development Studies. Trivandrum: Centre for Development Studies.

Purseglove, Jeremy. 1988. Taming The Flood: A History And Natural History Of Rivers And Wetlands. Oxford: Oxford University Press.

Radhakrishnan, P. 1981. Land Reforms in Theory and Practice. The Kerala Experience. *Economic and Political Weekly*, December 1981.

Radhakrishnan, P. 1981. Logan's Legacy. Economic and Political Weekly, July 11–18, 1981.

Radhakrishnan, P. 1982. Land Reforms and Changes in Land System. Study of a Kerala Village, *Economic and Political Weekly*, September 1982: A-107 to A-119.

Raja, Kunhan P. 1980. "Command Area" Its Concept, Objectives and Functions. In *Proceedings of Seminar on Water Management Practices in Kerala*. Calicut, 11–12, October 1980. CWRDM, Calicut.

Raju, K.V.; A. Narayanamoorthy; Govind Gopakumar; H.K. Amarnath. 2004. State of the Indian Farmer. A Millenium Study, Vol 3. New Delhi: Academic Foundation.

Ramachandran V K 2000. Kerala's Development Achievements and their Replicability. In *Kerala: The Development Experience. Reflections on Sustainability and Replicability*, ed. Govindan Parayil, pp. 88–115. London: Zed Books.

Ravi, S. P., C.G. Madhusoodhanan, A. Latha, S. Unnikrishnan, K.H. Amita Bachan. 2004. Tragedy of Commons. The Kerala Experience in River Linking. Thrissur: River Research Centre, New Delhi: SANDRP.

Reddy, Ratna V. 1999. Quenching the Thirst: The Cost of Water in Fragile Environments. *Development and Change*, 30(1999): 79–113.

Reddy, Ratna V. 2002. Water Security and Management. Lessons from South Africa. *Economic and Political Weekly*, July 13, 2002.

Rogers, Peter and Alan W Hall. 2003. Effective Water Governance. TEC Background Paper No. 7. Sweden: GWP.

Roy, Aditi Deb and Tushaar Shah. 2002. Socio-Ecology of Groundwater Irrigation in India. IWMI-TATA Water Policy Research Program.

Runge, Ford C. 1986. Common Property and Collective Action in Economic Development. *World Development*, 14(5): 623–635.

Saleth, Maria R. 2005. Water Institutions in India: Structure, Performance, and Change. In *Water Institutions: Policies, Performance and Prospects*, eds. Chennat Gopalakrishnan, Cecilia Tortajada and Asit K. Biswas, 47–80. Berlin, Heidelberg: Springer-Verlag.

Santhakumar, V and K. Narayanan Nair. 1999. Kerala's Agriculture. Trends and Prospects. In *Rethinking Development: Kerala's Development Experience*, ed. M.A. Oommen, pp. 314–324. New Delhi: Concept Publishing Company.

Santhakumar, V and R. Rajagopalan. 1993. Technological Prejudice: Case of Indian Irrigation. *Economic and Political Weekly*, April 3, 1993: 586–594.

Santhakumar, V., R Rajagopalan and S Ambirajan. 1995. Planning Kerala's Irrigation Projects. Technologial Prejudice and Politics of Hope. Economic and Political Weekly, March 25, 1995, pp. A-30-A-38.

Saradamoni, K. 1982. Women's Status in Changing Agrarian Relations. A Kerala Experience. *Economic and Political Weekly*, January 30, 1982: 155–162.

Schlager, Edella. 2006. Property Rights, Water and Conflict in the Western U.S. In *Changing Properties of Property*, eds. Benda-Beckmann F. von, K. von Benda-Beckmann Melanie Wiber, New York: Berghahn Books.

Schneider, Jurg. 1995. From Upland to Irrigated Rice. The Development of Wet-Rice Agriculture in Rejang Musi, Southwest-Sumatra. Berlin: Reimer.

Sengupta, Nirmal. 1996. The Indigenous Irrigation Organisation in South Bihar. In *Traditional Water Harvesting Systems. (An Ecological Economic Appraisal)*, ed. B. C. Barah, pp. 172–194. New Delhi: New Age International Limited.

Sengupta, Sohini. 2000. Political Economy of Irrigation: Tanks in Orissa, 1850–1996. *Economic and Political Weekly*. December 30, 2000.

Shah, Tushaar and K. Vengama Raju. 2001. Rethinking rehabilitation: socio-ecology of tanks in Rajasthan, north-west India. Water Policy, 3(2001): 521–536.

Shanmugaratnam, N. 1996. Nationalisation, Privatisation and the Dilemma of Common Property Management in Western Rajasthan. *Journal of Development Studies*, 33(2): 163–187.

Shiklomanov, I.A. 1997. Comprehensive Assessment of the Freshwater Resources of the World. Stockholm: World Meteorological Organization.

Sick, Deborah. 2007. Yours, Mine and Ours. Managing Water Resources in The Mexican-US Borderlands. In Baviskar, Amita (ed.) Waterscapes. The Cultural Politics of a Natural Resource. Uttaranchal: Permanent Black.

Singh, Chhatrapati. 1992. Water Law In India. New Delhi: Indian Law Institute.

Sivaramakrishnan, K. 1999. Modern Forests: Statemaking and Environmental Change in Colonial Eastern India. Stanford, CA: Stanford University Press.

Sneddon, Chris, Leila Harris, Radoslav Dimitrov and Uygar Ozesmi. 2002. Contested Waters: Conflict, Scale, and Sustainability in Aquatic Socioecological Systems. Society and Natural Resources, 15: 663–675.

Sokile, Charles and Barbara von Koppen. 2003. Local Water Rights and Local Water User Entities: the Unsung Heroines of Water Resource Management in Tanzania. Paper presented at the 4th WARFSA/WaterNet Symposium, 15–17 October, Gaborone, Botswana.

Sooryamoorthy, R. 2003. Fast Dying River Systems: Periyar and Bharathapuzha Basins. In *Managing Water and Water Users. Experiences from Kerala,* eds. Sooryamoothy, R and Antony Palackal S J, pp. 67–77. Maryland: University Press of America.

State Planning Board, 1975. Minor Irrigation in Kerala: An Evaluation Study. Trivandrum: State Planning Board.

State Planning Board, 1976. The Scheme for the free supply of pumpsets to Panchayats: An Evaluation Study. Trivandrum: State Planning Board.

State Planning Board, 1977. Intensive Paddy Development (Yela) Programme. An Evaluation Study. Trivandrum: State Planning Board.

State Planning Board, 1997. Ninth Five Year Plan (1997–2002). Report of the Task Force on Field Crops. Trivandrum: State Planning Board.

State Planning Board, 1998. Ninth Five Year Plan (1997–2002). Report of the Steering Committee on Water Resources. Trivandrum: State Planning Board.

State Planning Board, 1998. Ninth Five Year Plan 1997–2002. Report of the Steering Committee on Water Resources. Government of Kerala.

Tivy, Joy. 1990. Agricultural Ecology.

Tiwary, Rakesh. 2006. Explanations in Resource Inequality: Exploring Scheduled Caste Position in Water Access Structure. *International Journal of Rural Management*, 2(1): 85–106.

Todd, John. 1984. The Practice of Stewardship. In *Meeting the Expectations of the Land. Essays In Sustainable Agriculture and Stewardship*. Eds. Wes Jackson, Wendell Berry and Bruce Colman. San Francisco: North Point Press.

Torori, Cleophas O., Albert O. Mumma and Alison Field-Juma. 1996. Land tenure and water resources. In *In Land We Trust. Environment, Private Property and Constitutional Change*, eds. Juma, Calestous and J. B. Ojwang, Nairobi: Initiatives Publishers.

Turton, A. R.; B. Schreiner; J. Leestemaker. 2001. Feminization as a critical component of the changing hydrosocial contract. *Water Science and Technology*, 43(4): 155–63.

Turton, Anthony R. and Leif Ohlsson. 1999. Water Scarcity and Social Stability: Towards a Deeper Understanding of the Key Concepts Needed To Manage Water Scarcity in Developing Countries. SOAS Occasional Paper No. 17, Water Issues Study Group, London: University of London.

Ulluwishewa, Rohana. 1991. Soil Fertility Management of Paddy Fields by Traditional Farmers in the Dry Zone of Sri Lanka. *Journal of Sustainable Agriculture*, 1(3): 95–106.

United Nations Development Programme, 2006. Human Development Report, 2006. Beyond Scarcity: Power, poverty and the global water crisis.

Unni, Jeemol. 1983. Changes in the Cropping Pattern in Kerala. Some Evidence on Substitution of Coconut for Rice, 1960–61 to 1978–79. *Economic and Political Weekly*, September 1983.

Upadhyay, Videh. 2002. Water Management and Village Groups. Role of Law. *Economic and Political Weekly* December, 2002.

Van Koppen, Barbara, Charles S. Sokile, Nuhu Hatibu, Bruce A. Lankford, Henry Mahoo and Pius Z. Yanda. 2004. Formal Water Rights in Rural Tanzania: Deepening the Dichotomy?. Working Paper No. 71, Colombo, Sri Lanka: International Water Management Institute.

Vani, M. S. 2002. Customary Law and Modern Governance of Natural Resources in India-Conflicts, Prospects for Accord and Strategies. In *Legal Pluralism and Unofficial Law in Social, Economic and Political Development*, (ed.) Pradhan, Rajendra, Papers of the XIIIth International Congress on Folk Law and Legal Pluralism, Thailand, April 2002.

Vincent, Linden. 1990. The Politics of Water Scarcity: Irrigation And Water Supply In The Mountains Of The Yemen Republic. Network Paper 90/3e. London: ODI.

Vincent, Linden. 1995. Hill irrigation: water and development in mountain agriculture. London: Intermediate Technology Publications.

Vincent, Linden. 2003. Towards a smallholder hydrology for equitable and sustainable water management. Natural Resources Forum, 27(2): 108–116.

Vincent, Linden. 2004. Science, technology and agency in the development of drought prone areas: A cognitive history of drought and scarcity. PhD Thesis, Open University.

Vishwanathan, P.K. 2002. Irrigation and Agricultural Development in Kerala: An analysis of missed linkages. *Review of Development and Change*, 7(2) : 279–313.

von Oppen M., K V Subba Rao, and T. Engelhardt. 1986. Alternatives for Improving Small-Scale Irrigation Systems in Alfisol Watersheds in India. In *Irrigation Investment, technology and management strategies for development*, Easter, K.W. Studies in Water Policy and Management, No. 9, West View Press.

Vondal, Patricia J. 1987. The Common Swamplands of Southeastern Borneo: Multiple Use, Management, and Conflict. In *The Question of the Commons. The Culture and Ecology of Communal Resources*, eds. Bonnie J. McCay and James M. Acheson, pp. 231–249. Tucson: The University of Arizona Press.

Wade, Robert. 1988. Village republics: economic conditions for collective action in South India. Cambridge South Asian Studies No. 40, Cambridge: Cambridge University Press.

Wescoat, James L. Jr. 2002. Water Rights in South Asia and the United States: Comparative Perspectives, 1873–2000. In *Land, Property and the Environment*, ed. John F. Richards, 298–354. Oakland, California: Institute for Contemporary Studies.

Wilken, Gene C. 1974. Some Aspects of Resource Management by Traditional Farmers. In *Small Farm Agricultural Development Problems*, eds. H.H. Biggs and R.L. Tinnermeier, pp. 47–59. Fort Collins: Colorado State University Press.

Wilson, Kalpana. 2002. Small Cultivators in Bihar and 'New' Technology. Choice or Compulsion?. *Economic and Political Weekly*, March 30, 2002: 1229–1238.

Wolfe, Sarah and David B. Brooks. 2003. Water Scarcity: An alternative view and its implications for policy and capacity building. *Natural Resources Forum*, 27: 99–107.

Wood, Geof. 1999. Private Provision after Public Neglect: Bending Irrigation Markets in North Bihar. *Development and Change*, Vol. 30, 1999: 775–794.

Woodhouse, P. 2003. African Enclosures: A Default Mode of Development. *World Development.* , 31(10): 1705–1720.

World Commission on Dams, 2000. Dams And Development. A New Framework For Decision-Making. The Report of the World Commission On Dams. London: Earthscan Publications.